Professional Ideology & Development: International Perspectives

CANADIAN DEFENCE ACADEMY PRESS

Professional Ideology & Development: International Perspectives

Edited by
Lieutenant-Colonel Jeff Stouffer &
Justin C. Wright

CANADIAN DEFENCE ACADEMY PRESS

Canadian Defence Academy Press
PO Box 17000 Stn Forces
Kingston, Ontario   K7K 7B4

Produced for the Canadian Defence Academy Press
by 17 Wing Winnipeg Publishing Office.
WPO30380

Cover Photo: DGPA Combat Camera

Library and Archives Canada Cataloguing in Publication

Professional ideology & development : international perspectives / edited by Jeff Stouffer and Justin C. Wright.

Produced for the Canadian Defence Academy Press by 17 Wing Winnipeg Publishing Office.
Includes bibliographical references and index.
Issued by: Canadian Defence Academy.
ISBN 978-1-100-10315-0
Cat. no.: D2-231/2008E

1. Military art and science. 2. Armed Forces--Social aspects. 3. Military policy. 4. Sociology, Military. 5. Military education. I. Stouffer, Jeff II. Wright, Justin C III. Canadian Defence Academy IV. Canada. Canadian Armed Forces. Wing, 17.

U21.P76 2008   355   C2008-980302-7

Printed in Canada.

1 3 5 7 9 10 8 6 4 2

ACKNOWLEDGEMENTS

Professional Ideology and Development: International Perspectives represents the continued efforts and dedication of members of the International Military Leadership Association. As such, we would like to thank each of the contributing authors for their willingness to participate, professionalism, and sacrifice of their personal time to complete this volume.

The success of any publication, however, is also securely hinged on the professional technical support that is received throughout a project of this scope. To this end, we would like to express our fullest gratitude to Greg Moore of the Canadian Forces Leadership Institute for his timely and necessary production expertise. As always, 17 Wing Publishing Office, especially Captain Phil Dawes, Evelyn Falk and Michael Bodnar, are applauded for not only meeting project deadlines, but also for turning this volume into a finished and highly professional product. We would also like to thank Craig Mantle and Bill Bentley of the Canadian Forces Leadership Institute for their willingness to assist where required and for providing their professional subject matter expertise.

TABLE OF CONTENTS

FOREWORD

It is with great pleasure that I introduce *Professional Ideology and Development: International Perspectives*. This important and thought-provoking volume represents the third and latest effort of the International Military Leadership Association (IMLA), a recently formed academic organization whose purpose is to share military leadership knowledge and doctrine, as well as to identify opportunities for future collaborative research. At this time, the IMLA consists of representatives from various leadership institutes, research organizations and defence colleges from a host of diverse countries, including Switzerland, New Zealand, Singapore, Indonesia, the Netherlands and Canada. The continued commitment of the IMLA community to further collaborative understanding of leadership, professionalism and professional development is clearly reflected in this volume.

Within the following pages, current and diverse perspectives on the development of a robust military professional ideology are explored in considerable detail. As will be seen, in defining its professional ideology, each nation does so with a keen eye towards its own unique culture(s), history, politics, economics, resources and civil/military relations. To be sure, strong ideologies are never developed in a vacuum, and thus they are highly reflective of the society from which they spring. Regardless of the diversity of nations represented, each recognizes the fundamental requirement to develop and maintain a well-defined professional ideology, not only to function successfully in today's complex and ambiguous operating environment, but also to achieve a degree of credibility with, and attachment to, its respective society.

Of further interest, military practioners and like-minded academics have recently focused considerable attention on the need for militaries to expand their understanding of, and ability to contribute more effectively to, coalition and/or international operations. This book will undoubtedly figure prominently in discussions revolving around such issues because understanding the professional ideology of coalition partners will surely provide considerable insight into the many factors that will guide their subsequent behaviour and actions. The many challenges faced by the various countries represented in this book as they attempt to improve the capabilities of their military are also concisely articulated, as are the various "paths" that each have chosen to develop their own expression of professionalism in a military context.

Seen in this light, *Professional Ideology and Development: International Perspectives* stands as yet another significant accomplishment of the IMLA and is a testament to the value of collaboration across countries, and indeed, across

Foreword

continents. Should you wish to further discuss the contents of this volume – issues, challenges, opinions or future trends – please feel free to contact the individual contributors and/or the Canadian Defence Academy.

Major-General J.P.Y.D. Gosselin
Commander
Canadian Defence Academy

PREFACE

Events unfolding in the current international security environment have seemingly, once again, raised the spectre of an ongoing evolution in conflict, which many analysts and scholars argue is making global security even more chaotic and complex, if that is possible. Specifically, participation in international conflict is no longer necessarily restricted to clearly defined nation-states. In many cases, international terrorist and criminal networks, fuelled by ideological and religious extremism, as well as greed and the pursuit of power, have become major players. Moreover, the tactics are largely asymmetric in nature, with terrorism playing a major role. For many, the counter-thrust to this growing threat is increased co-operation and assistance. As a result, many nations have placed an emphasis on multinational military coalitions, coalitions of the willing, and the strengthening of international ties within existing organizations such as the United Nations.

Arguably, the ever-evolving security environment continues to change the way that most modern armed forces operate. There is a general recognition among military organizations of a need for its members to balance their tactical expertise and abilities with new skill sets that emphasize intellectual agility and practical adaptability in international and intercultural contexts. Parallel to these requirements are new sets of leadership challenges that stress continuous and progressive professional development across all levels of leadership and among all members of an armed force more generally. This implies an increasing emphasis on education, as opposed to a more traditional focus on training.

Part and parcel of the evolution in leadership and professional development currently being experienced by armed forces around the globe is the question of the development of their *professional ideology*. The professional ideology of an armed force is, generally speaking, the underlying set of principles that define its core identity and determine the manner in which its mission is achieved. The expression of these principles varies among nations and is specific or anchored to a number of factors such as culture(s), history, geography, economic prosperity, politics, social values and the beliefs of the nation. In essence, the professional ideology of an armed force is the expression of its institutional identity.

Given the prominence of professional ideology within any armed force, the question of development or, more succinctly, *change*, to that ideology becomes paramount. The need for transformation and new direction in leadership and professional development in the face of an ever-evolving and increasingly

Preface

complex security environment is clear. But meeting this requirement will also demand that armed forces strike a balance between flexibility and stewardship of their military identity.

Professional Ideology and Development: International Perspectives highlights the growing international recognition of the importance of continued leadership and professional development. This includes an awareness of the need to design and implement a comprehensive strategy to ensure continuous leadership development throughout their members' careers in the context of a new operating environment (both domestic and international) and recognizes the possibility of change to underlying professional ideology. This book outlines the various ongoing debates within the nations represented herein on the crucial role of professional ideology to the process of continued leadership and professional development. It speaks to the range of challenges that face militaries today in maintaining a balance between the need to adapt their professional development programs to maintain congruence with the values and beliefs of their host societies and their obligation to protect the professional ideology that underlies them.

Each nation represented in this volume faces its own unique challenges and culturally-specific issues within the ongoing debates around their professional ideology. For some militaries, even the use of the term "professional" in conjunction with their ideology or organizational identity is a subject of deliberation. For others, the definition of a "professional" military is perceived to conflict with the historical roots and socio-political status of an armed force among its nation's people and in the context of a still-developing nation.

For other militaries represented in this book, change to professional ideology is engulfed in a debate that focuses on the acknowledgement of gaps in operational effectiveness and leadership capacities, or the practical difficulty in supplementing or even moving away from tried and tested military knowledge and training and towards methods and mindsets yet to be vetted. Other challenges reported include the ongoing struggle with resource limitations in terms of funding, the availability of training capacities and facilities for program delivery, the increasing demands on personnel in light of the typically high operational tempo of current multinational operations, and the increased scrutiny of a global and highly interconnected media. The ongoing discourse in each of the nations represented in this book must remain fixed on their own unique priorities. Furthermore, any new direction or development to their professional ideology must, above all, reflect the role that its people wish their military to play.

Preface

Professional Ideology and Development: International Perspectives represents continued proof of the determined interest of the contributing nations in continuing to build international co-operation on both the study of leadership and professional development. It also speaks to the reality of the ways in which their unique professional ideologies must develop strategies and foster mutual understanding with their international allies. That is, the debates around changes to their professional ideology in response to the realities of today's security environment and their unique national priorities must also be embedded within the context of their willingness to contribute internationally.

In sum, *Professional Ideology and Development: International Perspectives* expands the body of leadership knowledge and emphasizes further collaboration. It aims to enhance the pursuit of leadership excellence by harnessing the contribution of the expertise and perspectives of other militaries, academic institutions and other key players involved with the International Military Leadership Association. It is expected that while reading this volume, readers will gain a better appreciation of the need for organizations to continue to develop comprehensive leadership and professional development strategies. Further, readers will recognize the rich potential for the advancement of leadership and professional development programs through the sharing of information, both new and tested, among military organizations. The contributors to and editors of this important book are commended for their continuing pursuit of excellence within the fraternity of the profession of arms.

Colonel Bernd Horn
Chairman,
CDA Press

CHAPTER 1

PROFESSIONAL IDEOLOGY IN THE CANADIAN FORCES

Dr. Bill Bentley

At the very core of any profession is the professional ideology that defines its special area of expertise and how that expertise is to be employed. This professional ideology stands in distinct opposition to the other two major ideologies extant in the socio-economic sphere of the Western capitalist market system, these are market ideology (entrepreneurialism) and bureaucratic ideology (managerialism). Professor Eliot Freidson contrasts these three ideologies in terms of how each views and organizes knowledge. In Freidson's terminology there are three "logics" underlying the nature, acquisition and application of knowledge in the areas of professions, the market and bureaucracies respectively. According to Friedson, in ideal-typical professionalism, specialized workers control their own work, while in the free market, consumers are in command, and in bureaucracy, managers dominate. "Each method has its own logic requiring different kinds of knowledge, organization, career, education and ideology."[1]

Military professional ideology is essential to the effective functioning of the profession of arms. It infuses and energizes the other three attributes of military professionalism – responsibility, expertise and identity – and embeds the profession in the society that it serves.

IDEOLOGY DEFINED

The concept of ideology, properly understood, is a powerful tool for analyzing how professionals view their role in society. The noted sociologist Talcott Parsons has argued that in the modern world, concern with the expression of moral commitments and with their application to practical problems, social and otherwise, has to a considerable degree become differentiated in the function of ideology.[2] Thus, for example, the ideology of professionalism asserts, above all else, devotion to the use of disciplined knowledge for the public good.

Ideology reflects a specific system of ideas or a conception of the world that is implicitly manifest in law, in economic activity and many other manifestations of individual and collective life. But it is more than a conception of the world as a system of ideas, for it also represents a capacity to inspire concrete attitudes and motivate action. To be recognized as such, an ideology must be capable of organizing humans; it must be able to translate itself into specific orientations for action. To this extent, ideology is socially pervasive or "the source of determined social actions."[3]

In his book *The Cultural Contradictions of Capitalism*, American sociologist Daniel Bell uses the concept of ideology by explaining that it is in the character of an ideology not only to reflect or justify an underlying reality, but once launched, to take on a life of its own. "A truly powerful ideology opens up a new vision of life to the imagination, but once formulated, it remains part of the moral repertoire to be drawn upon."[4]

The concept of ideology is therefore: systematic – beliefs about one topic are related to beliefs about another different topic; normative – to a large degree it contains beliefs about how the world ought to be; and, programmatic – it guides or incites concrete action.

PROFESSIONAL IDEOLOGY DEFINED

Systematic, normative and programmatic, professional ideology claims both a specialized, theory-based knowledge that is authoritative in both a functional and cognitive sense, and a commitment to a transcendental value that guides and adjudicates the way that knowledge is employed. It is functionally authoritative because the knowledge in question is the only knowledge that can get the job done. It is cognitively authoritative because it is ontologically grounded, theoretically-based and can only be fully accessed intellectually. The commitment at issue is represented in the ideology by the professional's occupational ethic, or in the case of the military profession, the military ethos. The ideology of professionalism, furthermore, argues that expertise properly warrants special influence in certain affairs because it is based on sustained systematic thought, investigation and experience, and in the case of individuals, accumulated experience performing work for which they had long and appropriate training and education.

THE KNOWLEDGE COMPONENT:
THE GENERAL SYSTEM OF WAR AND CONFLICT

The systematic, theory-based knowledge at the core of the military profession is the General System of War and Conflict comprising policy, strategy, operational art and tactics. This body of knowledge is illustrated in Figure 1.

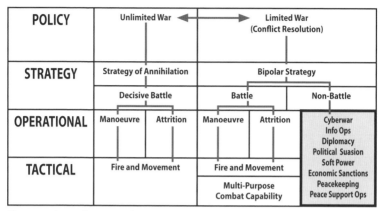

Figure 1: General System of War and Conflict.

In this model, war is defined as the continuation of policy with the admixture of other means. As the Prussian military theorist Carl von Clausewitz noted, warfare has its own grammar but not its own logic. In war, the logic is supplied by policy. Strategy, operational art and tactics constitute warfare, the grammar of war.[5]

Strategy is defined as the art of distributing and applying military means, or the threat of such action, to fulfill the ends of policy.[6] At the interface of policy and strategy lies the domain of civil-military relations and the zone in which all of the instruments of national power are aggregated and coordinated to generate a nation's overall national security strategy.

There are two fundamental types of strategy in the General System of War and Conflict model. If the political objective is unlimited or unconditional, then a strategy of annihilation is appropriate. Decisive military victory at the strategic level is the goal and only final victory counts. Both World War I (WWI) and World War II (WWII) provide good historical examples. If, however, the political goal is more limited, the bipolar strategy is the most appropriate choice. Here the strategist acts on both the battle pole and the non-battle pole, either simultaneously or sequentially. The Korean War and the Gulf War are good examples.

The operational level of war may be defined as the connecting link between strategy and tactics. It is at this level that campaign plans are created and executed. Operational art is the theory and practice of preparing and conducting operations (aggregated as campaigns) in order to connect tactical means to strategic ends.

Chapter 1

The tactical level of warfare is the realm of actual, direct combat. Here tactical manoeuvre – that is, fire and movement – is employed to physically achieve tasks assigned by the operational-level commander.

In the profession of arms, this core body of knowledge is augmented by a wide range of skills and other knowledge from a wide variety of disciplines (i.e., history, political science, the natural sciences, etc.). This supplementary knowledge is distributed and differentiated throughout the profession by rank and function.

THE ETHOS COMPONENT:
THE CANADIAN MILITARY ETHOS

Military ethos is defined as "the spirit that animates the profession of arms and underpins operational effectiveness." It acts as the centre of gravity of the military profession and establishes an ethical framework for the professional conduct of military operations.

In the Canadian context, the military ethos is intended to:

- establish the trust that must exist between the Canadian Forces and Canadian society;

- guide the development of military leaders who must exemplify the military ethos in their everyday actions;

- create and shape the desired military culture of the Canadian Forces;

- establish the basis for personnel policy and doctrine;

- enable professional self-regulation within the Canadian Forces; and,

- assist in identifying and resolving ethical challenges.

The Canadian military ethos is made up of three fundamental components: beliefs and expectations about military service; Canadian values; and, core Canadian military values.

Beliefs and Expectations

Beliefs and expectations of the Canadian military ethos include:

- **Accepting Unlimited Liability** – all members accept and understand that they are subject to being lawfully ordered into harm's way under conditions that could lead to the loss of their lives.

◆ **Fighting Spirit** – this requires that all members of the Canadian Forces be focused on and committed to the primacy of operations.

◆ **Discipline** – discipline helps build the cohesion that enables individuals and units to achieve objectives that could not be attained by military skills alone and allows compliance with the interests and goals of the military institution while instilling shared values and common standards. Discipline among professionals is primarily self-discipline.

◆ **Teamwork** – Teamwork builds cohesion, while individual talent and the skills of team members enhance versatility and flexibility in the execution of tasks.

Canadian Values

As a people, Canadians recognize a number of fundamental values that the nation aspires to reflect. Canadians believe that such values can be woven into the fabric of society. Canadian values are expressed first and foremost in founding legislation such as the *Constitution Act 1982* and the *Charter of Rights and Freedoms*. Other key values that affect all Canadians are anchored in other foundational legislation and articulated in their preambles.

Core Military Values

The third component of the Canadian military ethos (i.e., core military values) includes:

◆ **Duty** – obliges members to adhere to the rule of law while displaying dedication, initiative and discipline in the execution of tasks.

◆ **Loyalty** – entails personal allegiance to Canada and faithfulness to comrades; it must be reciprocal and based on mutual trust.

◆ **Integrity** – unconditional and steadfast commitment to a principled approach to meeting obligations. It calls for honesty, the avoidance of deception and adherence to high ethical standards.

◆ **Courage** – both physical and moral. Courage allows a person to disregard the cost of an action in terms of physical difficulty, risk, advancement or popularity.

The values, beliefs and expectations reflected in the Canadian military ethos are essential to operational effectiveness, but they also serve a more profound purpose. They constitute a style and manner of conducting military operations that earn for soldiers, sailors and air force personnel that highly regarded military quality – honour.

THE COMPETING IDEOLOGIES: MARKET AND BUREAUCRACY

Contrary to the professional's claim that only specialists who can do the work are able to evaluate and control it, both the ideology of the market and bureaucratic ideology claim a general kind of knowledge, superior to specialized expertise, that can direct and evaluate it.

Market ideology claims that ordinary human qualities informed by everyday knowledge and skills and fuelled by self-interest enable the individual, if properly motivated, to learn whatever is necessary to make all economic or political decisions. Denying the professional expertise any unique status, market ideology falls back on its own special kind of preparation for positions of leadership – an advanced but general form of education that they believe equips them to direct or lead specialists as well as ordinary citizens in the pursuit of profit.

The noted sociologist C.W. Mills warned of the dangers to professionalism posed by bureaucracy as early as 1956, arguing that bureaucracy was becoming such a dominant force in modern society that professions were increasingly "being sucked into administrative machines"[7] where knowledge is standardized and routinized into the bureaucratic apparatus and professionals become mere managers. The bureaucratic ideology claims the authority to command, organize, guide and supervise the activities of professionals. It denies authority to professional expertise by claiming a form of knowledge that is superior to specialization because it can organize it rationally and efficiently. Those who espouse this view of knowledge from within the ranks of bureaucracy or management could be called "elite generalists."

STEWARDSHIP OF THE PROFESSION

Ensuring that professional standards are established and strengthened is certainly one of the most important roles for institutional leaders in the Canadian Forces (CF). Continuous execution of this responsibility is essential to enhancing operational effectiveness and is the primary means of protecting the profession from the insidious encroachments of both market ideology and especially bureaucratic ideology. Stewardship must focus on both components of professional ideology – the maintenance and growth of the core body of knowledge at the heart of the profession and the promulgation of the military ethos to align military culture with its dictates. Given the inclusive nature of the profession of arms in Canada, stewardship is the responsibility of both officers and senior non-commissioned members (NCMs) serving in the regular and reserve force. As such, stewardship is defined as the special obligation of officers and non-commissioned members who by virtue of their rank or appointment are directly concerned with ensuring that the profession of arms in Canada fulfils its organizational and professional responsibilities to the Canadian Forces and Canada. This includes the use of their power and influence to ensure the continued development of the profession, its culture and its future leaders to meet the expectations of Canadians.

CONCLUSION

The military profession is always under a degree of pressure that threatens to erode its basic tenets and characteristics. These threats include those posed by the competing ideologies discussed above, as well as societal changes that may be anathema to true professionalism. In addition, new roles and responsibilities in the uncertain security environment of the 21st century must be accommodated. This requires new forms of expertise while retaining a clear sense of identity as members of the profession of arms. It is the role of professional ideology, properly nurtured, to bind the profession firmly and to ensure its future vitality.

ENDNOTES

1. Eliot Freidson, *Professionalism* (Chicago: University of Chicago Press, 2001), 6.

2. Talcott Parsons, "The Professions," *Encyclopedia of the Social Sciences*. (NY: Macmillan, 1937), 545.

3. Jorge Larrain, *The Concept of Ideology* (Athens, GA: University of Georgia Press, 1979), 80.

4. Daniel Bell, *The Cultural Contradictions of Capitalism* (London: Heineman, 1976), 60.

5. Carl von Clausewitz, *On War,* in Michael Howard and Peter Paret, eds. (Princeton: Princeton University Press, 1976), 146. See also General Sir Rupert Smith, *The Utility of Force* (London: Allen Lane, 2005).

6. *Leadership in the Canadian Forces: Leading the Institution* (Ottawa: DND, 2005), 42.

7. C.W. Mills, *White Collar* (NY: Oxford University Press, 1956), 112.

CHAPTER 2

THE SINGAPORE ARMED FORCES CORE VALUES: THE BUILDING OF A SHARED IDENTITY AND PROFESSIONAL ETHOS

Lieutenant-Colonel Kim-Yin Chan, PhD
Kwee-Hoon Lim
Major Adrian Y.L. Chan
*Colonel Sukhmohinder Singh**

INTRODUCTION

When one first joins or interacts with any organization, a natural starting point is to ask for the corporate mission, vision and values of the organization in order to better understand it. However, while such explicitly espoused "corporate" values are often part of a branding, marketing or image-building exercise, they are not necessarily a reflection of the actual values and ideals that guide the everyday behaviour of most people in that organization.

The Singapore Armed Forces (SAF) Core Values did not originate from "corporate branding" interests. Rather, the Core Values emerged as part of a journey that began with attempts to clarify civil-military relations when the SAF first implemented National Service in 1967. This eventually culminated in a readiness to articulate more explicitly the "SAF's Character" in the 1990s through a set of Core Values.[1] This marked a coming of age for the SAF as a military institution, as many well-established military forces also articulate their Core Values, not just as a corporate branding effort, but as a statement of professional-organizational ideals and principles to guide individual and sub-unit actions.

CHALLENGES OF FORGING A SHARED IDENTITY

When Singapore achieved independence in 1965, many doubted that the city-state would survive. The abruptness of its birth as a nation meant that

---

*   The views expressed in this chapter are those of the authors and do not necessarily reflect the official policy or position of the Singapore Armed Forces, the SAF Centre of Leadership Development, SAFTI Military Institute, the Ministry of Defence, or the Singapore Government. The authors would like to thank Mr. Kuldip Singh, Head of Centre for Heritage Services, MINDEF, for vetting this chapter and for his valuable comments. The authors are also grateful to the early pioneers of the Singapore Armed Forces for laying a firm foundation for the building of the SAF Core Values and the forging of a shared identity and professional ethos.

a unified national identity was not something that it could count on as the social glue for the nation's diversified population base (i.e., Chinese and Indian immigrants, European ex-colonial masters, and native Malays). Indeed, inter-racial tensions underpinned the early years of the nation, culminating in the race riots of the 1960s.

Against this backdrop of an uncertain nationhood and a fragile national identity, Singapore needed to quickly expand its military standing force of two infantry regiments to assume the duties of defence from British forces, who had signalled their intention to pull out of Singapore by 1971. To assist with this military build-up, Singapore enlisted the help of Israeli defence advisors.[2] In 1967, two years after independence, compulsory national service was introduced. This initiative was challenging to implement since it entailed the involuntary enlistment of men from diverse backgrounds to be trained under the supervision of foreign defence advisors. At the same time, conscription was also culturally counter-intuitive for the ethnic Chinese who comprised the majority of the enforced conscription. For them, joining the military was culturally unattractive because "only the useless son will be a soldier...they didn't see soldiering as an achievement."[3]

Given the above context, the challenges of forging a shared identity for the SAF were formidable. For example, the young military had no history and limited to no traction with the populace, particularly the ethnic Chinese. Military culture was alien to the conscripts and they were generally unwilling to be a part of it. What little tradition was inherited from the British military was not always in sync with the new methods and doctrine being introduced by the Israeli advisors. In addition, with the primacy of rapid military expansion to increase the size of the standing force, there was limited opportunity available to develop the softer, yet critical, aspects of shared identity and professional ethos.

THE EARLY YEARS:
FROM CODE OF CONDUCT TO SAF CORE VALUES

From the onset, the SAF realized that it could not simply depend on inherited traditions from the British to anchor its identity and professional ethos. Rather, it had to forge its own code of conduct taking into consideration its unique milieu. In particular, its social standing in the young Singaporean society needed to be addressed. For some time, the Singapore community viewed soldiers as little more than "jagas" (local slang for security guard).[4] Hence, the SAF Code of Conduct (COC) was created and formally declared on 14 July 1967 in preparation for the enlistment of the first intake of national servicemen in August 1967. The COC emphasised the unique purpose and identity of the SAF: "Members of the SAF have a unique role – they are not only the ever vigilant guardians of our nation but are also required to be an example of good citizenship. The COC is the foundation of character, conduct and discipline required of every member of the SAF."[5] During a press interview given at the enlistment ceremony, Defence Minister Dr. Goh Keng Swee explained the rationale for the COC: professional efficiency and the need to clarify the relation between the armed forces and society.

Without a strong military tradition or history of successful campaigns, the SAF decided that it should anchor itself on the unique purpose of the military and the basics of the military profession – values and competence. Defence Minister Dr. Goh Keng Swee captured this in a speech at the promotion ceremony of a group of senior SAF officers in 1972:

> ...A military elite differs from other kinds of elite in a number of respects. First their function in society is obviously different. Military elites are the ultimate guardians of the independence of sovereign states. They ensure the independence of nations by their ability to deter or resist military aggression and absorption by another sovereign state. ...military elite [sic] place regard on the nature of values such as honour, loyalty, physical courage, professional pride, distaste for luxury, contempt for wealth, liking for physical life and so on. ...such values in military such values in miliraty elite in general come about either as a result of successful military campaigns, or during periods of peace, and are actively cultivated in the life of professional soldiers. You should take note of these values and try to foster them not only among yourselves but also to inculcate them in your junior officers. ...pride in your profession or in your unit or in yourself should not be a superficial one, like the pride of a peacock in its resplendent plumage. Rather, it should be the result of proven achievement, mastery over techniques and thorough knowledge of military subjects. In other words, pride in one's profession should be the result of professional competence.[6]

Two years later, in an internal work plan speech, Dr. Goh Keng Swee further reinforced this realization and the commitment to establish a uniquely SAF ethos as one founded on the "profession of arms":

> In the history of men, various armies at various times have adopted different methods of cultivating a high level of esprit de corps among its officer corps. In Singapore, we try to emulate the British and in this instance, with notable lack of success. The British Army, especially

the infantry arm, cultivates pride, loyalty and comradeship among its officers within the framework of the regimental tradition. Officers derive inspiration from the history of the regiment which is physically embodied in the regimental colours emblazoned with its battle honours. In everyday work, an active social life centring around the regimental mess fosters camaraderie. We have tried to transplant these practices in our army. I have come to the conclusion that they do not work, and possibly cannot be made to work. Indeed, it would be astonishing if it were otherwise, seeing that not only are our military systems different, but also that we are two different peoples, with different histories, customs, social values, individual perceptions and group responses. We will have to find our own methods of fostering esprit in the officer corps, which will fit into our own social environment as well as our systems of military organization. I do not believe that this can be achieved by resorting to gimmicks; it will be a long term and long haul effort over many years. Whatever methods ultimately evolve, the basic ingredient of military esprit de corps remains the same in all armies and through all ages. This is pride in one's calling – the profession of arms. Without such pride, it is not possible to develop a community spirit. But if esprit de corps were to be a reality not a show-off, professional pride of this kind must rest on the solid base of professional competence.[7]

While such professional pride and ethos was encouraged from the onset of the SAF's development, they were not formalized in doctrine. Indeed, the lag between concept articulation and formal doctrine was considerable. For example, early speeches by political leaders made reference to "Leadership by Example" as a philosophy desired of SAF leaders. Mr. Lim Kim San, the Defence Minister in 1970, stated at a commissioning parade: "...in our concept, to command is not only to order but to lead."[8] But it was only in 1984 that "Leadership by Example" and "People-oriented Management" were formally adopted as the SAF's leadership and management philosophies respectively, as part of a document called *The SAF Declaration*.[9] In some ways, this doctrinal lag reflects the readiness of the SAF to embrace the softer aspects of military maturation.

As the SAF continued to build on its basic organization, design, technologies and operating concepts for greater efficiency and effectiveness, leadership development received a huge boost in the late 1980s with a project initiated by then-Prime Minister Lee Kuan Yew for the SAF to create an "Institute of Excellence" – namely, the present Singapore Armed Forces

Training Institute – Military Institute (SAFTI-MI). In 1987, a project team tasked to envision this "Institute of Excellence" proposed the need for an Officer's Creed as a central thesis to the concept of SAFTI-MI. The project team also proposed that the SAF needed to articulate a set of core values as a basis for guiding the nature of SAF leadership. The idea was to identify a common set of core values such that those trained under the system would bear distinctive and positive attributes that are uniquely SAF. These core values would apply to all ranks in the SAF, however, a deliberate decision was made to gradually "cascade" the values down the ranks commencing with the officer corps. Hence, in 1990, the *Officer's Creed* was launched, coinciding with a fundamental change in the design of the officer cadet course that featured a tri-service and professional element in place of the previous curriculum with a generic junior-senior term dichotomy.

With the redesigned officer-training curriculum, reinforced by the *Officer's Creed*, officer cadet training became more centred on a formal articulated ethos. After several cohorts of officers had graduated from this new curriculum, in 1996, a decision was made to fully promulgate the SAF Core Values to all members of the SAF. Hence, from 1997, a booklet entitled The SAF *Core Values: Our Common Identity* (also known as the 1997 onwards, *SAF Core Values Handbook*) was published by SAFTI Military Institute for distribution to all newly enlisted recruits in the SAF.

SAF CORE VALUES AS A FOUNDATION TO THE SAF'S CHARACTER AND IDENTITY

In his Foreword to the 1997 SAF Core Values Handbook, the then-Chief of Defence Force Lieutenant-General Bey Soo Khiang wrote: "The SAF Core Values should be synonymous with the SAF Character. When one thinks of the SAF and a SAF soldier, the qualities associated with the organization and its members should exemplify prominently the SAF Core Values."[10] The SAF Core Values underpin the professional "character or identity" of all SAF personnel. The 1997 Handbook stated that the SAF Core Values were meant to act as "the foundation upon which a quality armed forces is built," to "shape our professional beliefs and attitude" and "bind our people together in fulfilling our professional roles and duties."[11] It went on to say: "This set of shared values is also important in meeting the demands and challenges of war. It serves as a guide in the way we conduct ourselves as professionals."[12] It is therefore important to note that the SAF Core Values are not simply civilian, corporate-organizational values; they represent an articulation of the professional, military identity that is important for the SAF's specific mission context. Hence, to build the character of the SAF, one must focus on developing the SAF Core Values in each and every member of the SAF.

WHY THESE SEVEN CORE VALUES?

The seven SAF Core Values are:

- *Fighting Spirit;*

- *Professionalism;*

- *Discipline;*

- *Ethical Conduct;*

- *Leadership;*

- *Loyalty to Country;* and,

- *Care for Soldiers.*

These seven Core Values were chosen after much thought – they are not civilian values that could be applied to any corporate organization. Rather, these values were chosen for their utility to the SAF. Indeed, the original classified documents that debated and proposed the values included arguments for each of the values that were very specific to the mission and organizational context of the SAF. For example, *"Fighting Spirit"* was justified from lessons learned during the rapid fall of Singapore during the Second World War. As a small island-nation without any strategic depth, it was vital to instill in SAF soldiers a fighting spirit that was to be ferocious, able to take or withstand the initial blows and to stand back up and fight.

On the other hand, *"Professionalism"* was justified on the basis that although the SAF is largely a conscript force, the SAF soldier is "not a civilian in uniform". A SAF soldier is expected to be highly proficient and reliable in all his military tasks, his conscript nature notwithstanding.

"Discipline" is about military discipline (i.e., the responsible obedience of orders and the timely and accurate execution of tasks). Discipline is necessary if military units are to overcome the severe stress resulting from the temptations, confusions and pressures of combat. It demands a respect for and appreciation of the military system and the role the military plays in defence of the nation.

"Ethics" was included because it was absolutely essential for SAF troops to uphold the international laws of armed conflict and to be responsible in the use of military force. Entrusted with instruments of force, SAF troops have a special moral obligation to temper the use of force with a sense of ethical conduct. A strong offensive spirit, if given free rein, might lead to acts that violate international morality.

Interestingly, *"Leadership"* was justified for inclusion as a SAF Core Value because of a concern that technological advances and the increased "civil-ianisation" of the SAF through a gradual devolvement of supporting organizational processes to civilian entities might lead to commanders becoming increasingly "managerial" and thus losing their leadership edge. "Leadership is what rises to dominate the battlefield."[13]   Hence, leadership must always remain a defining characteristic of the SAF.

The next Core Value, *"Loyalty to Country,"* was articulated to capture the idea that military service begins with a responsibility and commitment as citizens to protect and defend the nation. The extent of this duty is such that it may require SAF soldiers, sailors and airmen to sacrifice their lives for Singapore. For a conscript armed force, being loyal also includes willingness to commit one's time, talents and energy to serve the SAF.

The final Core Value, *"Care for Soldiers,"* was adopted in 1997 on the basis of the SAF being largely a conscript force. It is therefore vital that the regulars never forget the need to care for the soldiers entrusted to them. "Care" includes training the soldiers as realistically as possible so that they would survive in war and caring for them as individuals while they serve their nation.

In sum, the SAF Core Values are comprised of *professional*, *moral* and *leadership* components. The professional component includes *"Professionalism,"* *"Discipline"* and *"Fighting Spirit."* The moral component includes *"Loyalty to Country,"* *"Ethics"* and *"Care for Soldiers."* *"Leadership"* is the core value under the leadership component.

RELEVANCE OF THE SAF CORE VALUES
FOR A 3RD GENERATION SAF

In 2004, the concept of a "Third Generation" or "3G SAF" was announced in the Singapore Parliament. "3G SAF" became the rallying call to guide the transformation of the SAF in response to changes in the operating environment. These transformations included the need to be able to execute

Chapter 2

an expanding spectrum of military operations, to better absorb and leverage national service enlistees who are increasingly more educated and technologically-savvy, and to better integrate the use of military technology.

In the same year, the SAF Centre of Leadership Development (SAF CLD) was tasked to examine the relevance of the SAF Core Values for the 3rd Generation SAF and to propose a strategy for the way forward. The SAF CLD revisited the roots of the SAF Core Values to examine the purpose and intent of each of the values and ascertain their relevance to the future context.

The study concluded that the Core Values were still relevant in the 3rd Generation SAF operating context, if not even more so. For example, the rapid spread of the Severe Acute Respiratory Syndrome (SARS) virus in 2003, and the plot by the extremist group Jemiah Islamiah to plant bombs at various locations in Singapore soon after September 11th 2001, gave new importance to *"Fighting Spirit"* beyond its traditional link to ferocity in defence of limited physical-strategic depth. Beyond being prepared, fighting spirit is important to help one sustain the initial blows from terrorists or diseases because they can strike unexpectedly and from any direction.

With globalization, individual citizens have greater access and opportunities to work and settle in other countries. While increased mobility is important for economic growth, it may also potentially dilute commitment and loyalty to nation. In this sense, *"Loyalty to Country"* and *"Ethics"* have become even more important in the 3rd Generation conscript-based SAF. Similarly, advances in information technology now grant greater autonomy to individuals regarding information control and dissemination. The pervasiveness of media presence in operations, and the sense of autonomy afforded by what is commonly termed as New Media, increasingly demand that soldiers exercise discretion and self-regulation. In this regard, ensuring national security through information prudence takes on greater significance in this new environment. Hence, the notions of *"Loyalty to Country"* and *"Ethics"* now also include the sense of responsibility to the nation in terms of how soldiers today handle potentially sensitive information.

The pervasive use of advanced military technology means that military decision-making cycles are more intense, requiring greater precision and synchronization across dispersed decision nodes. This in turn compels corresponding changes in behavioural capacities for effective social networking. One such capacity is *"Discipline"* in a network – an individual or small unit "node" in a system must play its part to "synchronize" with changes in the network environment. With the greater dispersion of forces afforded by advances in communications technology, commanders must not lose sight of the importance of both *"Leadership"* and *"Care"* for their soldiers.

EARLY EFFORTS TO INCULCATE THE SAF CORE VALUES

In a monograph entitled *Spirit and Systems: Leadership Development for a Third Generation SAF*, it was noted that, "Just as Singapore spent its first 10 years building a shared Singaporean identity, the first 10 years of the SAF Core Values program has succeeded reasonably well in shaping a common identity across the three services and across the ranks in the SAF."[14] To be more precise, the early success hinged on getting the Core Values to the rank and file. These early efforts, beginning with the dissemination of the 1997 SAF Core Values Handbook, focused on communicating the meaning and importance of the Core Values to the masses via direct means such as communications posters and talks, as well as through appropriate role modelling of the Core Values by leaders and instructors. These early efforts have been hugely successful. Today, the Core Values are effectively 10 years old in a 40-year-old SAF. It is difficult to find any SAF personnel who do not know of the SAF Core Values.

On the other hand, beyond communication, there was a wide variation in how deeply the Core Values took root. Some units invested beyond communication requirements by assigning secondary appointments to their officers to explore "values inculcation" in the units, while others stopped at just rote memorization of the Core Values by its rank and file. By and large, it is fair to say that the early years did not produce any consistent doctrine on how to inculcate values.

A NEW STRATEGY FOR CORE VALUES IN THE SAF

The focus of the early years was to promulgate the values through a prescriptive, "explaining" approach supported by leadership role modelling. However, the downside to such an approach was a tendency to act only as prescribed by rules and guidelines – where the rules were silent, the behaviours were agnostic of the spirit and intent of the Core Values. In this sense, there was a need for a new approach to internalize the Core Values to serve as one's own moral compass.[15]

Given this need, the SAF CLD embarked on a new initiative to make sense of existing Core Values education efforts through understanding "military ethics education" in other forces. This led the SAF CLD to develop an initial "theory" of the key steps in values education that was followed by a broad strategic framework to guide future application of the SAF Core Values in the organization, one which should better embed the Core Values in the culture of the SAF. In 2007, a formal strategy was articulated to move the SAF's agenda for Core Values beyond merely communicating or explaining, to a broader effort to embed the Core Values in the culture of the SAF. To achieve this, four reinforcing "domains of action" were identified to organize interventions at different levels of effects in the organization (Figure 1). The four domains of action are:

- **(SAF) Character-building domain.** These refer to interventions that strengthen the professional military identity of individuals in the SAF. Such interventions include socialization practices, clarification and alignment of personal versus SAF values, etc.

- **(SAF Values-based) Choice-making.** These refer to ethical reasoning interventions that strengthen the ability of SAF individuals and teams to make decisions on the basis of the Core Values. Such interventions include education and training in ethical reasoning, the conduct of after action reviews based on the SAF Core Values, etc.

- **(SAF Values-based) Organizational/Unit Climate.** These refer to the extent to which the SAF Core Values exist as the basis of the ethical climate in the organization and its units.

- **(SAF Values-based) Communities.** These refer to the extent that the SAF Core Values are the basis of social networking across different functional domains of the organization (e.g., alumni, societies/clubs, communities of practice, etc.).

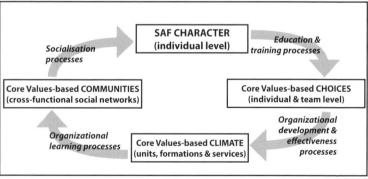

Figure 1: Four Reinforcing Domains of Action to Embed Core Values in SAF Culture.

The strategy acts as a way to organize the efforts of centralized, educational/ training establishments (i.e., focusing on character-building and choice-making) and decentralized, service and formation-led initiatives that act to strengthen the ethical climate and build values-based professional communities in (and out) of the SAF. We have also identified organizational learning tools and practices, (e.g., story-telling, values clarification, ethical decision-making, command team-building, and community-of-practice knowledge-sharing), which are most effectively applied at different segments of the Framework (Figure 2).

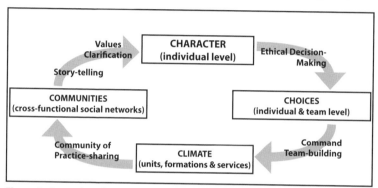

Figure 2: Examples of Methods and Practices to be Infused to better Embed Core Values into SAF Culture.

EXECUTING THE STRATEGY

Having developed a long-term strategy for embedding the Core Values into the culture, the next challenge is to implement the interventions at the different Domains of Action. The SAF CLD decided that the key starting point for the strategy was at the foundation military courses, including the three-month Basic Military Training Course, the nine-month Officer Cadet Course and the six-month Section Leader Course.

In addition, three things were identified to support the interventions. First, to be systematic about intervention implementation, there is a need to situate a theory of values internalization for individuals within the SAF context. Second, reliable and valid instruments are needed to measure the effect of the interventions as well as the longer-term SAF-level outcomes. Finally, there is also a need to profile the "baseline" attitudes toward the SAF Core Values amongst the full-time National Servicemen (NSF) and regulars. Of these, the first effort undertaken by SAF CLD in 2007 involved the development of measures of "baseline" attitudes toward the SAF Core Values.

Chapter 2

THE NEED FOR A THEORY OF VALUES
INTERNALIZATION FOR THE SAF

The unique make-up of soldiers in the SAF presents significant challenges and prevents the wholesale adoption of existing theories of value internalization without careful consideration of the SAF context. For example, the SAF is comprised of young adults who are legally obligated to serve two years of national service at age 18. Of these, about 10% move on to be further trained as junior commanders (up to platoon level) and who eventually become responsible for role modelling and inculcating values in subordinates not much younger than themselves. This means that any value internalization efforts for these NSF commanders need to be sharp and effective enough for them to quickly become socialized, so that in turn, they can become agents of cultural change for their peers.

These NSFs move into the "reserve" as Operationally Ready National Servicemen (ORNSmen), ready for operational call-up duties until age 55. During this phase, their values continue to develop outside the direct influence of SAF organizational socialization. For them, there is a need to ensure continued alignment between their personal values, organizational values from their civilian jobs and the SAF Core Values so that they can be quickly integrated into the SAF during their call-up for ORNS duties.

Finally, the SAF is also comprised of mostly regulars who signed on with the SAF at age 18. The legal age for enlistment is 18 years. There were a few exceptional cases where people enlisted earlier with parental consent. Some women join the forces after their "O" levels (Ordinary Levels General Cambridge Examinations – students take this exam after completing four to five years of secondary level education). These regulars then serve for up to 20-25 years before transiting to a second career outside the SAF. For them, opportunities exist to fully experience organizational socialization. However, at any point in time, the age of SAF regulars can range from 18 to 45, which means that these regulars are differentially socialized to the organization. Any value internalization effort will need to take this diversity into account. In particular, the older regulars (aged 35 and above) will have experienced a different educational pedagogy from the NSFs that they are leading. This too needs to be taken into account.

Given these unique challenges, it is unlikely that any one theory of values internalization will be sufficient to serve the diverse needs of these four distinct population profiles (i.e., NSFs, ORNSmen, regulars, and different age cohorts). Nonetheless, common to all is the need to achieve a degree of alignment between personal and organizational values that takes into account the individual's current level of personal values clarity. With these broad starting points, work is under way to develop theories of values internalization for the SAF.

"LEVELS OF INTERVENTION" THEORY

The remainder of this chapter presents our efforts and progress regarding values inculcation in the SAF. Here, we present the theory and progress of our efforts to embed the SAF Core Values into the culture of the SAF.

Our theory proposes that efforts to inculcate organizational or professional values in employees can be seen in terms of interventions that take place at three levels: clarifying and aligning of personal with organizational values in the context of the organization's mission and purpose; understanding the "threats to the values or to values-based actions"; and, making choices among the values through a process of ethical reasoning. This "theory" is illustrated in Figure 3.

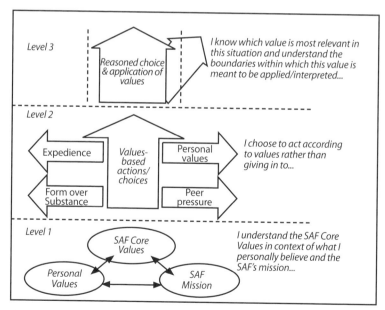

Figure 3: A Theory of Values Inculcation Involving Three "Levels of Intervention."

Chapter 2

ALIGNING PERSONAL AND ORGANIZATIONAL VALUES

For individuals to embrace organizational core values, it is important that they have an appreciation of their personal values. To facilitate acceptance of these organizational core values by the individual, these values need to resonate or be aligned with the individual's personal values. Even then, there are various degrees of internalization. On the one hand, one can simply *accommodate* the organizational core values as additions to one's personal values. At the other end of the continuum, one can *assimilate* these core values into one's personal values such that one's personal values set is expanded or enriched (i.e., better *differentiated* or better *integrated*). Regardless of the degree of eventual acceptance, when introducing these organizational values to a recruit's personal values, one should expect the average soldier to initially be resistant to these additional values, unless his/her personal values are highly aligned to those of the organization. The greater the gap between the personal values and the organizational values being introduced, the stronger will be the initial resistance. Hence, it is important to know the degree of alignment of the personal values to the organizational values being introduced. This is illustrated in Figure 4 below.

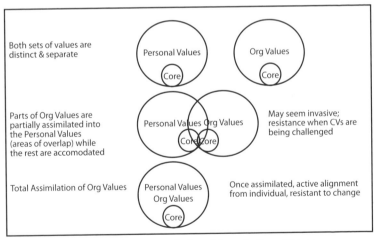

Figure 4: Assimilation and Accommodation of Organizational Values.

PROCESSES AND TOOLS FOR VALUES INTERNALIZATION

Figure 5 below depicts the relationship between personal and organizational values clarification, the alignment between the two, and their subsequent applications in choice-making. It also presents some tools we have found useful at each stage to facilitate the internalization of organizational values (OV) to one's existing personal values (PV) set.

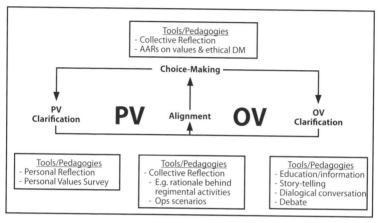

Figure 5: Processes and Tools for Values Internalization.

The starting point of values internalization is the clarification of one's personal values. Here, a short survey of values can be administered to help one become more self-aware of one's personal values. The purpose of self-reflection at this stage is also to facilitate self-awareness. Next, organizational values such as the SAF Core Values can be introduced via various methods, ranging from general information giving to more creative methods such as story-telling in order to articulate the nuances of the values.

SOME CAVEATS

There are several caveats that need to be highlighted. First, there is some degree of individual difference as to how organizational values are received. Some automatically seek alignment. Others choose to reflect before deciding. Yet others may eventually choose not to align. Because of the conscript nature of the SAF, it is highly likely that all three types of responses will be found in our enlistees. This should be expected and differentially dealt with by commanders.

Second, it is important to recognize that alignment is not automatic and cannot be forced. Forced alignment is often counter-productive. Nonetheless, it is also important to note that even when one is not yet aligned to organizational values, it does not mean that they will not execute values-influenced behaviours. It is more important that any form of values-influenced behaviour be encouraged as a starting point, even if the values being demonstrated are personal values rather than the desired organizational values. By and large, it is easier to draw linkages between personal values and organizational values when personal values are strong. Hence, values-based behaviours are to be encouraged.

Third, the military operating context compels quick conformity and the demonstration of desired behaviours. Such behaviours include demonstrations of regimentation (linked to discipline), safety, professionalism, sacrifice (linked to loyalty to country), and fighting spirit. Hence, it can be argued that for new conscripts, conformity and demonstration of desired behaviours are probably more critical than waiting for them to manifest such behaviours through internalization. What is important to be aware of is that when new conscripts demonstrate such behaviours, one should not be deceived into believing that alignment through internalization has occurred. This is critical, because conformity is achieved during peacetime at the expense of expending command and control. In situations when command and control are weak, for example during the fog of war, conformity cannot be relied upon to compel the same desired behaviours in soldiers. Under such situations, one's internal moral compass takes over. If the moral compass is composed of personal values, then the soldier will act for his or her own benefit. If the moral compass is composed of the organization's desired values, then the soldier will act as a responsible member of the organization.

Finally, while conformity is not a desired end state, it is nonetheless a useful intermediate step for values internalization. Conformity over time leads to overlearning, which is simply habituation of behaviours. When habituation occurs, behaviours are demonstrated automatically. What needs to be recognized, however, is that conformity is not a useful intermediate step for some types of values-related skills and capacities, such as moral awareness and reasoning. For these types of skills and capacities, one needs to revert back to either selection or development through scenario-based training and discussions.

CONCLUSION

Like a child growing into adulthood, one's identity is ever-evolving and ever-developing. Yet it retains elements of permanence that is attributed as "character." Compared to other siblings in the same family, the child retains a unique identity, yet shares some similarities with the siblings that are attributed to "family resemblance." The same can be said of how the character of the SAF is developing.

In this chapter, we have attempted to trace the SAF's journey in building its identity against the backdrop of Singapore's own emerging nationhood. We articulated how and why the seven Core Values were chosen by the SAF, the challenges faced by the SAF in its early years of values inculcation, as well as the next steps to embed the SAF Core Values into the culture of the SAF. As the operating environment and mission profile of the SAF changed, the SAF adapted itself and yet at the same time retained its essential elements – the character of the SAF. As the SAF progresses into the next phase of embedding the Core Values into its culture, each and every soldier will work towards achieving a shared "family resemblance" that marks each soldier as unique and yet belonging to the family of the SAF.

ENDNOTES

1. Kim-Yin Chan, Sukhmohinder Singh, Regena Ramaya and Kwee-Hoon Lim, *Spirit & System: Leadership Development for a Third Generation SAF [4th Pointer Monograph]* (Singapore: SAFTI Military Institute, 2005).

2. *The Singapore Armed Forces,* Martin Choo, Hock-Sen Chan, Robert Chew & Kheng-Joo Low, eds. (Singapore: Public Affairs Department, Ministry of Defence, 1981), 19.

3. Anthony Kang, "Enlisted for National Service in 1967," in Carl Skadian, Psalm Lew, Bryan Wong, Emen Low, Minghui Fong, Jason Lim, Keith Lim, Sean Lim, Matheus Yoe, Choon-Ping Ng and Melvyn Ong, eds., *40/40: 40 Years & 40 Stories of National Service* (Singapore, Ministry of Defence: Landmark Books Pte Ltd., 2007), 49.

4. Speech by Defence Minister Dr. Goh Keng Swee at the Armed Forces Day Parade at Jalan Besar Stadium on 1 July 1971.

5. Statement by Defence Minister Dr. Goh Keng Swee at the Press Interview on public declaration of the SAF Code of Conduct on 14 July 1967.

6. Speech by Defence Minister Dr. Goh Keng Swee at the Promotion Ceremony for Senior SAF Officers at MINDEF HQ on 2 May 1972.

7. Speech by Defence Minister Dr Goh Keng Swee on the Introduction of the MINDEF Workplan 1974/75 for the SAF at the SCSC Auditorium on 7 March 1974.

8. Speech by Mr. Lim Kim San, Minister for Defence, at the Commissioning Parade at SAFTI on 14 February 1970.

9. Chan et al., *Spirit & System*, 13.

10. *The SAF Core Values: Our Common Identity*, A SAFTI Military Institute Publication, 1997, i.

11. Ibid., 2.

12. Ibid., 3.

13. "A Value System for the Institute of Excellence", an internal staff paper by SAF Joint Operations Department, dated 23 October 1987.

14. Chan et al., *Spirit & System*, 57.

15. Ibid.

CHAPTER 3

A PROFESSIONAL IDEOLOGY FOR THE NETHERLANDS ARMED FORCES

Lieutenant-Colonel drs. Coen van den Berg
Dr. Elly Broos
Lieutenant-Commander drs. Ineke Dekker-Prinsen
Major drs. Robbert Dankers
*Lieutenant-Colonel drs. Dick Muijskens**

The Netherlands Armed Forces is an all-volunteer professional armed force that is firmly tied to Dutch society, whose vital interests it functions to serve. Further, it contributes to international peace and security, often under challenging circumstances. At the same time, the Netherlands Armed Forces is a professional organization with a rational model for managing its performance and remains accountable to the Dutch government. For this reason, the Dutch Armed Forces has (since 2005) used an organizational model with central paradigms that focus on strong central steering of results, internal billing for products and services, and a separation between policy-making and operations. However, it is expected that Dutch soldiers can perform their job under life-threatening circumstances where the use of violence may be needed and that they are able to cope with the hardship of deployments (be it peacekeeping, combat or humanitarian operations, or a combination of these). Being a soldier, however, has become much more than just being a fighter. Dutch soldiers are now obliged to carry out their duties in a professional, proper and responsible way. As such, the Netherlands military is confronted with the necessity to re-formulate its professional ideology. For example, some of the domains in which soldiers' professionalism is challenged and where questions concerning proper and responsible conduct are faced include: dealing with violence; cross-cultural collision; tight budgets; recruitment of youth; different types of missions (human relief, peace enforcement); and, shifting professional ideology due to societal changes. Increasingly, the Netherlands professional soldier is confronted with ethical dilemmas in his/her working environment.

---

*   Disclaimer - The views expressed in this chapter are those of the authors and not the Netherlands Armed Forces.

Chapter 3

Recently, discussions over professional ideology in the Netherlands Armed Forces have been provoked by events concerning sexual harassment. As an example, considerable political attention was drawn after an incident that occurred onboard a navy vessel. An investigation by the "Staal Commission" led to the conclusion that uniformed organizations, like the armed forces, show a higher incidence of such immoral behaviour compared to civilian organizations.[1] It was subsequently made very clear by the leadership in the Netherlands Armed Forces that any occurrence of immoral behaviour would be considered a violation of the spirit of the Netherlands military. Moreover, it was concluded that commanders need to deal with such incidents, not just in a procedural legal way, but also in a way that ensures the care of the personnel under their command. In particular, this conclusion led to new discussion that emphasized the need to formulate a clear vision on leadership for the military profession. The results of the "Staal Commission" led to the development of two internal documents that largely focus on how to carry out the military profession in a proper and responsible way: the Behavioural Code of the Netherlands Armed Forces,[2] hereafter named the "code of conduct," and the Integral Vision on Leadership,[3] hereafter named the "leadership vision." These documents bypass the wider discussion on how the military has to formulate its professional ideology in terms of the wider range of challenges confronted by soldiers today. These documents, however, fall somewhat short in that they do not deal with the image of the military professional outside the military eventhough the military always operates within a wider social context and hence could not exist in isolation from it. We argue that a professional ideology is necessary for the military profession but also for Netherlands society in general, since society provides the basis for the existence and legitimacy of the military profession.

The above-mentioned policy documents were intended to stimulate desired behaviour but were instead perceived by a large number of soldiers as a denial of their existing professional attitude and behaviour. They believed that they acted and behaved in accordance with these principles long before these documents were produced. Although the overall content of these documents was not contested, they were perceived, especially with the code of conduct, as an unfounded criticism of their responsibility and professionalism. In particular, the code of conduct has limited itself to a set of general statements:

- I am part of a professional organization;
- I am a member of a team with a common task;
- I am aware of my responsibility;

- I am trustworthy and treat everyone with respect; and,

- I ensure a safe working environment.

We argue that these statements apply to professional behaviour in general instead of focusing specifically on the underlying values of the military profession.

Professional ideology in the Netherlands Armed Forces has in our opinion the imperative to lead to more self-consciousness and pride in the military profession. We argue the need for this from two perspectives. First, clarity about the military professional ideology will provide soldiers with a clear vision about the important values of their profession. It could thus inspire a sense of pride and honour. Second, society will have increased insight into what it is that makes the military profession unique, valuable and necessary. A clear professional ideology will anchor the military profession in Netherlands society and serve to clarify the important role the military plays in the smooth functioning of modern society. It is our belief that this will also contribute to more effective recruitment for the armed forces. But what should the content of professional ideology be for the Netherlands Armed Forces?

The code of conduct and the leadership vision make explicit reference to the field of work of the military and mention aspects like deployment and working under adverse circumstances. In the Netherlands Defence Doctrine, leadership is the projection of the personality and the character of the individual, mostly the commander, to motivate soldiers to do what is expected of them.[4] In addition, the leadership vision asserts that personnel are the most important asset of the armed forces. The Netherlands vision of effective leadership is based on the pillars mentioned in the code of conduct, namely, professionalism, teamwork, responsibility, respect, integrity, and safety. The underling values are also mentioned explicitly in the leadership vision. These are honesty and transparency, courage, empathy and partnership, service-ability, internal commitment, and authenticity. These values, which are explained in the next section, lead to character building and enclose the adagio at the gates of the Netherlands Defence Academy *"Kennis is macht, karakter is meer"* [Knowledge is power, character is more].

Honesty and transparency. Leaders explain clearly to their staff what is expected of them and why. Leaders have a clear vision and transfer this vision. Leaders act soundly, are an example for their staff and are able to give feedback effectively, thereby respecting their employees.

Chapter 3

Courage. Leaders are prepared to accept risk in our missions (in administrative or commercial surroundings, as well as in dangerous operational circumstances). Leaders represent their staff and share a willingness to demonstrate their vulnerability when necessary. Leaders express their opinion clearly but with the respect for others. Leaders are prepared to learn from mistakes.

Empathy and partnership. Leaders are sincerely interested in others and listen to their staff. They give subordinates full attention and appreciation, and show trust in them. Leaders stimulate the personal development of their staff and ensure a safe working environment. Leaders work actively to achieve a strong and diverse team.

Serviceability. Leaders are available to serve the team and the task. They delegate responsibility and provide opportunities to their staff to develop themselves and learn from their mistakes. Leaders are flexible.

Internal commitment. Leaders radiate energy and optimism and take the initiative. Leaders believe in the defence organization passionately, want to share this passion with others and carry the passion forward.

Authenticity. Leaders are able to deal with disappointment. They have insight into their own qualities and are willing to improve their weaknesses. Leaders confront challenges and difficult situations and are willing to develop themselves.

The Royal Netherlands Navy, Army, Air Force and Military Police (MP) have each separately prepared a document in which they further explain their leadership values. In an internal document from the Royal Netherlands Navy,[5] a number of leadership competencies were described: co-operation, integrity, responsibility, capacity to learn, communication, courage, flexibility, delegating, interpersonal sensitivity and the development of staff. When the descriptions of those competencies are compared to those described in the leadership vision, many similarities can be observed. The same applies to the leadership visions of the Royal Netherlands Military Police[6] and the Royal Netherlands Army,[7] while the credo "every soldier a rifleman" introduced by General Van Uhm in his position as Commander-in-Chief of the Royal Netherlands Army adds to the reality of a soldier's profession.[8] In the army, however, the relationship between ethics and the military profession is emphasized extensively. The air force vision on leadership is very much influenced by the technical and safety aspects of our weapon systems and a focus on the end state. "One team, one mission" with a central emphasis on promoting peace and security. Good co-operation between leader and subordinate, to reach determined goals, is therefore of extreme importance. Core values in this vision include co-operation, clear goals, model conduct, reciprocal trust and respect.[9] In order to believe in the defence organization as passionately as is described in the leadership vision, it is essential to have a clear image of what the defence organization is and what it stands for. Using this vision and the resulting central values, the requirement for an explicit professional ideology for Netherlands Armed Forces personnel is discussed, as are current developments in this field. Some of these developments include the status of the military profession and its contribution to society, the development of military science and the demand for quality over lifelong education.

THE STATUS OF THE MILITARY PROFESSION: CONTRIBUTION TO SOCIETY

As explained, the Netherlands Armed Forces vision on leadership and the code of conduct were created in response to undesirable behaviour in the defence organization. Unfortunately, these documents are primarily inwardly focused and applicable largely to employees currently working in the organization. In our opinion, however, it was necessary to formulate a professional ideology for the Netherlands Armed Forces that also fosters recognition of the military profession in the wider society. The military profession needs to be accepted by society as an honourable profession and a career of choice. Fortunately, the image of the Netherlands Armed Forces remains relatively positive. Support for military operations, however, is not automatic and is prone to change as witnessed by the declining public support for the present mission in Afghanistan.[10] In light of serious recruitment and retention problems in the Netherlands Armed Forces, the articulation of what a soldier stands for has become increasingly important.

In addition, the nature of the military profession has moral implications; seemingly contradicting terms like "enforcing peace" illustrate this point. The military has a mandate for the proper and responsible use of violence, thus in certain situations a choice needs to be made between life and death. With this in mind, the moral dilemmas of the soldier should not turn into personal dilemmas.[11] It is our contention, however, that soldiers in the Netherlands Armed Forces likely have a somewhat different experience as demonstrated by the increase in postponed Post-Traumatic Stress Syndrome amongst soldiers having participated in military combat situations.[12] The dilemmas are not strictly specific to the military profession as they also have a link to the public character or essence in that they concern the quality of the service of the soldier as a representative of Netherlands society. Explicit and open forms of military ethics reinforce the capacity to test and justify choices made by soldiers in ethical dilemmas in contrast to implicit and introverted forms of military ethics.[13] The consequence of this for the Netherlands Armed Forces is the need to communicate in a clear and transparent manner with society.

A professional ideology could serve to help achieve clarity for both civilian and military personnel with respect to what a soldier stands for in terms of standards, values and morality. At present, an ideology has not been clearly defined nor made operational, the obvious result being the persistence of multiple interpretations of ever-present ethical dilemmas in the context of the "fog of peacekeeping missions." This is undesirable since public opinion and support are essential requirements for the legitimacy of the military profession in the Netherlands. The Second World War and Cold War clearly marked the need for a military profession. The fall of the Berlin Wall (9th November 1989), however, somewhat clouded the need for a professional ideology. Moreover, as a result of the downsizing of the military after the Cold War, the Netherlands transformed its armed forces into an all-professional force and abolished conscription in 1996. Though smaller in size, recruiting was best described as enlisting vast numbers of soldiers in the lower ranks with most only staying for a limited period of time. The public image of the armed forces and especially the image of the military profession as a (temporary) career choice became the focus of considerable attention. This resulted in tension over recruitment campaigns in which the methods and reasons to attract people became quite diverse, ranging from contributing to peace and security, to adventure, to acquiring job experience, as well as campaigns that provided realistic information about the essential role of the military profession in Dutch society. After recent terrorist attacks directed at western societies, especially the 9/11 attacks in New York (11th September 2001) and the attacks in Madrid (11th March 2004), the need for a military profession as well as the demands on the military profession have become increasingly clear. The Netherlands Armed Forces has gradually become more involved in missions with a higher potential for danger (i.e., Afghanistan).

The current constitutional tasks of the Netherlands Armed Forces are threefold: to defend national and allied soil; to contribute to international peace and security; and, to provide assistance to national and international authorities with humanitarian aid. Those tasks emphasize the underlying values of respect and dignity for human life, values that are closely bound to the values of society and international (humanitarian) law.

Since the military profession has a mandate for the use of violence, a high moral standard is essential. As such, each soldier carries the responsibility to effectively deal with ethical dilemmas. At present, ethics within the Netherlands Armed Forces is primarily focused on how to deal with peace-keeping and humanitarian situations. Military ethics are primarily related to the use of violence by the armed forces during military operations, and as such, are different from dealing with violence in civilian situations. The use of violence that is inherently present during military operations is subject to democratic opinion and is controlled by law. As such, it is surrounded by legal and normative ethical principles that are accepted in the Netherlands. The complexity of the military profession lies primarily in the paradox of being an armed force that must ensure peace and security through the means of violence.[14] The fact, however, is that military ethics has become essential for current military tasks and operations. Furthermore, this implies that each soldier needs to conduct him/herself professionally at all times and remains ultimately responsible for his/her own actions. Moral awareness, hence, is included in the training of all soldiers. The challenge is to bridge the apparent gap between the normative concept in the sense of defining ethical behaviour and the contextual concept whereby the various conditions that influence ethical behaviour are included in such training.[15] Research in the Netherlands Armed Forces has shown that typical military ideals and values like obedience, honour and national pride are receiving less attention in the initial training of young officers and that values like honesty, responsibility, making a contribution to peace and security, as well as respect, have taken their place.[16] As such, it is important to include ethical behaviour as a normative concept that is founded in a system of values and virtues that belong to the military profession. Hence, moral competencies have to be defined in a military professional ideology for the Netherlands Armed Forces.

Besides the code of conduct, international views also influence the development of a military professional ideology in the Netherlands. The code of a warrior, as defined by Shannon French,[17] for example, not only defines how the soldier should interact with his/her comrades, but also how he/she should treat other members of his/her society, the enemy and the people he/she conquers.[18] French further argues that such a code restrains the soldier and provides behavioural boundaries. In doing so, the code of a warrior distinguishes an honourable act from a shameful act. Important questions in this regard are: how can a soldier understand and internalize the meaning of words like sacrifice, honour and courage? Does a soldier have clarity concerning his/her moral and legal duties and responsibilities as well as to whom he/she is accountable?[19] When such questions are answered and individual commitment is ensured, it influences the behaviour of the soldiers, and as such, it could make the difference between respect and rejection by society.

Psychological research shows that, especially under threat, a soldier's commitment to the group and the leader,[20] as well as commitment to a professional ideology, are important factors that influence his/her behaviour. Ethical dilemmas have become intertwined in the military profession and it is not by training and preparation alone that a soldier debates his or her responsibilities and accountability for his/her actions. As potential dilemmas concerning the use and interpretation of violence are considered during mission preparation, it is in the public arena that the justification, norms and values of the armed forces and the wider civil society are debated. The realization that this belongs to the military profession was stated by one of the Dutch battle group commanders of the Stabilization Force in Iraq in 2004. In an interview, a journalist asked whether investigations by the Dutch Public Prosecution Service on legal aspects of the use of force by Dutch troops would influence soldiers' work. The response:

> Of course this will do so to some extent, but my men are eager to do their job. We know that we have to account for situations in which violence has been used. It is never pleasant when the process of accountability leads to the Court of Justice. But we do not need to fear this, as long as one acts carefully. Former detachments have been confronted with three such cases and in all cases the Dutch Public Prosecution Service ordained that the use of force had been within the restrictions of the Rules of Engagement.[21]

Incidents also occur in the Netherlands Armed Forces as a result of competing or dual professions where clashes are sometimes experienced between individuals' respective professional values. A military doctor, for example, who was in charge of managing medical provisions while participating with the Dutch battalion (Dutchbat) during a peacekeeping mission in Srebrenica (former Yugoslavia) was faced with a dilemma.[22] In this instance, two types of provisions required management: the regular provision that could be used for humanitarian aid, and, the "iron stock", which was exclusively reserved by military superiors for Netherlands soldiers (taking care of their own personnel was seen as the primary responsibility). After the fall of Srebrenica, a civilian became seriously injured, however, the regular provision was depleted (i.e., it had been used for general humanitarian purposes). In such a situation, is the Hippocratic Oath or medical declaration of Geneva in direct violation of the vision or intent of military superiors?

Current operating environments, characterized with high stakes on military, political and civilian fronts, further heighten the need for military personnel to have sound expertise in a number of areas.[23] For example, it is essential to show competencies such as cultural sensitivity and the ability to build meaningful relationships with military as well as civilian organizations/professionals within the Netherlands and internationally. One could argue that a professional ideology in this sense has the imperative to function as the moral justification for acceptance as an actor in the field of international stability and security. In this sense, cultural sensitivity is an essential competency required by the military.

A SOLDIER'S CAREER IN DEMANDING AND CHANGING SURROUNDINGS

In 2008, a more Flexible Personnel System (FPS) was introduced to enhance the ability of the Netherlands Armed Forces to continue to effectively participate in international or coalition operations.[24] The aim of the FPS is to ensure that the right quality and quantity of service personnel are available and to allow service personnel to work in different armed forces branches if they have the requisite competencies. The FPS also identifies specific moments during a soldier's career where transitioning to civil society is most appropriate. As this transition is an explicit element of a military career, a soldier's developed professional competencies have to be valuable to help ensure that he/she is qualified and competent to enter the civilian sector. This mutual interest of service personnel in the military profession and personnel in the wider society also demands that the military express its professional ideology. To do so, the military profession has to express itself on the following five fields: development of military science; status of the military profession and military contribution; demand for education, skills, and experience; continued education; and development of a clear professional ideology that is based on a coherent collection of ideas, views and notions concerning the military profession. The following sections address current developments in each of these areas in the Netherlands Armed Forces.

THE DEVELOPMENT OF MILITARY SCIENCE

The development of military science is very dynamic at present. The scientific education of officers at the Royal Netherlands Military Academy and the Royal Netherlands Naval Institute has always known a duality of being scientific or theoretical while instilling the right military attitude. The collaboration of the Netherlands academies and institutes for various levels of officer education resulted in a Faculty of Military Science in the Netherlands

Chapter 3

Defence Academy in 2005. This has given military science a new élan. The Netherlands Defence Academy is also in the process of gaining recognition as a university in line with the Bologna declaration of 1999. This process should result, by the end of 2008, in the recognition and subsequent awarding of bachelor degrees for officer graduates of the Netherlands Defence Academy. This process goes hand in hand with the development of more in-depth military science training and possibly a graduate studies program. While some of the ambiguousness of military attitudes towards these developments remain, a growing number of officers are participating in PhD studies. As present operational demands have taught us, current operations require officers who can cope with complex problems. As such, the need for academic skills is acknowledged among a growing number of Netherlands officers and accreditation of the Netherlands Defence Academy by other universities in 2008 would be a positive step to support this development.[25]

STATUS OF THE MILITARY PROFESSION AND MILITARY CONTRIBUTION

The leadership vision is a very strong instrument to enhance corporate identity and foster a clear expectation on how to organize, prepare and lead troops into action. When carried out, any task or mission should not confront personnel with any surprise or leave personnel uncertain as to what course of action is expected. A clear vision allows personnel to understand inner thoughts, beliefs and expectations from any leader in the organization. An important part of that vision involves our central values. For any task or action, one can ask if it is carried out in concert with the central values. Colonel M.K. Leboeuf of the US Army stated this as follows:

> ...Organizational vision is most similar to the Army concept of *commander's intent*. ...*Operations* describes commander's intent as the description of operational purpose and end state. *Purpose* is what you want the organization to do – it is the single unifying focus of the operation. *Method* is generally how you intend for your subordinates to carry out their jobs. *End state* is what you want the final result to be.[26]

The internal communication of this vision is of the utmost importance. Any means of communicating, be it through email, official meetings, posters, and especially through the setting of a good example, are needed to clearly define organizational values and leadership vision. Since 2005, command over the Netherlands Armed Forces was centralized to one Chief of Defence, General Berlijn. From the outset, he communicated his leadership vision; it was clear and concise. Every month he wrote an intranet column in which he addressed past and present experiences. He also related each of these experiences to our professional values. Trust, pride, professionalism, serviceability, courage and respect appeared in every other sentence. His written communications were concluded with a reminder to his troops about the honour he has had in being able to co-operate with them. Besides writing about leadership, General Berlijn was a role model in that he lived his leadership vision. A clear and troops-supported leadership vision is an important element to successfully understand and complete specific tasks in peacetime situations. This vision is also equally if not more important in difficult surroundings when troops are disorientated and separated from their leaders. Thus, when the common purpose, the methodology and the end state are well communicated and understood, and when personnel are committed to the professional values of the organization, it is easier to empower them to achieve mission success.

The fall-out or consequence of a clear and supported leadership vision is how personnel represent or discuss the Netherlands Armed Forces with family, friends, international partners and the media. Further, we have to be transparent about our (planned) operations and carry out our promises. This is how society will perceive the Netherlands Armed Forces and how it will determine what status to give us within our Kingdom. This is also an important factor to use during recruitment and the training of young soldiers. The aforementioned FPS attracts more and younger personnel with certain expectations about the organization. An interesting instrument in the recruitment process is the search for and acceptance of previously acquired competencies (*Eerder Verworven Competenties* (EVC)) to meet the much needed human resources requirements in the Netherlands Armed Forces. In this sense, the education, training and experience of new personnel should directly recognize the expected culture, status, and professional ideology of the organization. This speeds the process of enculturation with a well-based, collective behaviour being the result. This also provides an element of stability in an organization that is consistently challenged by political policy changes as well as the new and never before experienced conflicts in unknown parts of the world.

THE DEMAND FOR EDUCATION, SKILLS AND EXPERIENCE

As soldiers are drawn from society for complex tasks, the demand for education and relevant skills has become increasingly important. It must be recognized, however, that military personnel will at some point transit back to civil life and, in most cases, another career. The FPS makes explicit the need for civil education, skills and experience for the armed forces. It also recognizes that military education, skill and experience can and should be transferable to the civilian sector, and as such, represents an essential prerequisite for recruitment and retention strategies.[27] By implementing a system that recognizes previously acquired competencies, the Netherlands Armed Forces has increased its potential for recruitment among Dutch youth, especially among those who do not fit prescribed entry qualifications, but have acquired essential competencies through life or civilian job experiences. Moreover, recognizing competencies acquired during one's military career adds to the value of former soldiers when entering the civilian labour force and bolsters the professional value and image of soldiers as professionals. Both for the development of military and other competencies needed to fulfill one's job as a soldier or member of society, the two-way recognition of competencies (i.e., military and civilian) will be crucial for the flexibility and effectiveness of the Netherlands Armed Forces.

THE DEMAND FOR QUALITY AND LIFELONG EDUCATION AND MORAL PROFESSIONALISM

As the demand for professionalism can be made explicit through education, skills acquisition and experience, lifelong education has become paramount in today's rapidly changing world. In this regard, learning how to learn and redefining mental models have become important aspects for individuals and groups in organizations.[28] In general, it could be argued that learning processes within an organization need to adjust to the new context of an organization and its visionary objectives.[29] Thus, soldiers have to keep up with developments both within the organization as well as in society. In addition, it is argued in the context of the Netherlands Armed Forces that military professionalism includes moral professionalism.[30] This means that the military professional ideology as well as the underlying values are frequently discussed and debated and that soldiers are willing to redefine their mental models and resulting behaviour in accordance with such an ideology.[31]

THE DEVELOPMENT OF A CLEAR PROFESSIONAL IDEOLOGY

For further development of a professional ideology for the Netherlands Armed Forces, theoretical models, the outcome of dialogue with soldiers and citizens in the wider social surroundings and lessons learned from their professional experience can be used. As already mentioned, a coherent collection of ideas, views and notions concerning the military profession can be used to develop a clear professional ideology. The real challenge, however, is the integration of lessons learned, debates, dialogue, discussion leading to new ideas, views and notions concerning the military profession into a professional ideology, including its transference and acceptance to the military community as a whole. As previously stated, the Behavioural Code of Conduct and Integral Vision on Leadership, as well as the current developments in the FPS, are seen as the instruments to guide the implementation of the professional ideology. As a result of the Staal report, military leaders are trained to focus on the social aspects of leadership and as such the focus on well-being and organizational climate play an important role. Both in career courses as well as in the assessment of leaders, these aspects are addressed. Coaching, counselling and intervention are additional instruments that are used for this purpose. Attention must be paid to these processes, as every soldier has to fit within the moral values of the framework of the professional ideology.

Moreover, as every soldier is seen as a professional, regardless of rank, they have to fit into the system of acquiring specific competencies (EVC). Essential to this process is the commitment to the professional ideology and its consequences. However, some processes that are essential for the realization of behaviour consistent with the professional ideology have to be embedded in teams or units. For this purpose, intensive team-building sessions are part of pre-deployment mission preparation.

CONCLUSION

Writing on this topic has helped to elucidate elements of ideas, views and notions Dutch soldiers have of their profession that are implicitly or explicitly related to a Dutch military professional ideology. At the same time, we have realized that the Dutch military professional ideology has not been brought together in one clear articulation. We therefore see it as a challenge to contribute to the discussion and to formulate a professional ideology for the Netherlands Armed Forces. The fields that we have identified for such an ideology include: continued development of military science; status of the military profession and contribution; demand for education, skills and experience; continued education; and development of clear professional ideology that is based on a coherent collection of ideas, views and notions concerning the military profession. We believe it is important to have a strong attachment to moral competencies in such an ideology, as the essence of the Netherlands Armed Forces is to defend the core values of Dutch society. It has to do so under possibly life-threatening circumstances where high demands are made on human performance. Likewise, the Netherlands Armed Forces is an organization that is responsible for over 60,000 service members that require a safe and secure work place and working conditions. As such we argue that a military professional ideology can only exist within the framework of a flexible, effective and accountable organization.

Chapter 3

ENDNOTES

1. B. Staal, H. Borghouts and J. Meyer, *Ongewenst gedrag binnen de krijgsmacht. Rapportage over vorm en incidentie van en verklarende factoren voor ongewenst gedrag binnen de Nederlandse krijgsmacht* [Undesirable Behaviour within the Armed Forces. Report Concerning Form and Incidence as well as Explanatory Factors for Undesirable Behaviour within the Dutch Armed Forces], 2006, 1–6. Retrieved 15 March 2008 from http://www.mindef.nl/binaries/Rapport%20Ongewenst%20Gedrag%20binnen%20de%20Krijgsmacht_tcm15-67260.pdf.

2. *Gedragscode Defensie* [Code of Behaviour] (Ministerie van Defensie: Den Haag, 2007).

3. *Visie Leidinggeven* [Leadership Vision] (Ministerie van Defensie: Den Haag, 2008).

4. *Nederlandse Defensie Doctrine* [Dutch Defence Doctrine] (Defensiestaf: Den Haag, 2005).

5. Koninklijke Marine, *Taakgericht en mensgericht leiderschap in balans. De visie op leiderschap binnen de Koninklijke Marine* [Finding a Balance in Task- and People-Directed Leadership. The Leadership Vision within the Royal Navy] (Royal Netherlands Navy: Den Helder, 2007).

6. Koninklijke Marechaussee, *Leiderschap met lef* [Leadership with Bravados] (Royal Netherlands Military Police: Den Haag, 2007).

7. Koninklijke Landmacht, *Handboek Leidinggeven Koninklijke Landmacht* [Handbook for Leaders in the Royal Army] (Royal Netherlands Army: Den Haag, 2002).

8. Generaal Van Uhm, Integer en Doelgericht [Sound and Purposeful] (Dag, 18-4-2008: Netherlands newspaper). Retrieved 12 June 2008 from http://www.dag.nl/1070438/Nieuws/Artikelpagina-Nieuws/Profiel-Van-Uhm-Integer-en-doelgericht.htm.

9. Koninklijke Luchtmacht, *Kijk op leidinggeven Beleidsvisie* [Vision on Leading] (Royal Netherlands Airforce: Den Haag, 2004).

10. Interview with minister Van Middelkoop. Retrieved 22 April 2008 from http://www.nos.nl/nos/artikelen/2008/04/art000001C8A3C90BA8B6C6.html.

11. Adriannus Iersel and Theodore Baarda, *Militaire ethiek. Morele dilemma's van militairen in theorie en praktijk* [Military Ethics. Moral Dilemma of Soldiers in Theory and Practice] (Budel: DAMON, 2002).

12. G. Smid, T. Mooren, R. Van der Mast, R. Kleber and B. Gersons, *Stresstoornis: systematische review, meta-analyse en metaregressieanalyse van prospectieve studie* [Stress Disorder: Systematic Review, Meta-analysis and Metaregression Analysis of a Prospective Study] (Tijdschrift voor Psychiatrie, 2008).

13. Iersel and Baarda, *Militaire ethiek.* [Military Ethics].

14. Adriannus Iersel, Theodore Baarda and Desiree Verweij, *Praktijkboek militaire ethiek. Ethische vraagstukken, morele vorming, dilemmatraining* [Practice Book Military Ethics. Ethical Questions, Moral Shaping, Dilemma Training] (Budel: DAMON, 2004).

15. William Kahn, "Toward an Agenda for Business Ethics Research," in Gary Yukl, ed., *Leadership in Organizations (6th ed.)* (New Jersey: Prentice Hall, 2006).

16. Desiree Verwey, "Morele professionaliteit in de militaire praktijk" [Moral Professionalism in Military Practice], in J. Kole and D. de Ruyter, eds., *Werkzame Idealen* (Assen: Van Gorcum, 2007).

17. Shannon French, *The Code of a Warrior* (Lanham: Rowman & Littlefield Publishers, 2003).

18. Ibid.

19. Ibid.

20. Paul Bartone and Faris Kirkland, "Optimal Leadership in Small Army Units," in R. Gal and D. Magelsdorff, eds., *Handbook of Military Psychology* (Chichester: John Wiley & Sons, 1991).

21. LtCol Kees Mathijssen [Interview with commander battlegroup SFIR4 in newspaper] (Rotterdam, NRC Handelsblad, 24-7-2004).

22. Iersel, Baarda and Verweij, *Praktijkboek militaire ethiek.* [Practice Book Military Ethics]. Adrianne van Es and Vincent de Jong, "Mensenrechten en gezondheidszorg," Retrieved 17 May 2008 from http://www.johannes-wier.nl/userfiles/file/Cursusboekjuni2001.pdf.

23. Peter Kiestra and Pieter Simpelaar, *Leiderschap met lef* [Leadership with Bravados] (Koninklijke Marechaussee, 2006).

24. HDP/Directie Personeelsbeleid, *Flexibel Personeelsysteem* [Department of Personnel Policy, Ministery of Defence, Flexible System of Personnel Management] (Internal document: Den Haag: OBT bv., 2007).

25. S.J. (Julian) Lindley-French and Anne Tjepkema, "Krijgsmacht x wetenschap = krijgswetenschap," *[Armed forces * sciens= military science], Militaire Spectator,* 177(5), (2008), 286-294.

26. Maureen Leboeuf, "Developing a Leadership Philosophy," Retrieved 17 May 2008 from www.gcsc.army.mil/milrev./english/mayjun99/leboeuf.htm.

27. Visie EVC binnen de Defensieorganisatie [Vision on Previously Acquired Competencies with the Defence Organization] (Internal document: Ministerie van Defensie, Nota, 25-4-2008).

28. Yukl, *Leadership in Organizations.*

29. Albert Kamperman, "Leidinggeven en HRM" [Leadership and HRM] in F. Kluytmans, ed., *Leerboek Personeels Management* (Groningen: Wolters-Noordhoff, 2005), 347–371.

30. Desiree Verwey, An Introduction NIME at a Leadership and Ethics Symposium 27 May 2008 at Soesterberg.

31. Desiree Verwey, "Morele vragen en dilemma's in de militaire praktijk" [Moral Questions and Dilemmas in the Military Practice], in M. Mentzel, ed., *Filosofie & Praktijk: Oorlog en Vrede* [Philosophy & Practice: War and Peace] (Budel: DAMON, 2001).

CHAPTER 4

PEOPLE'S ARMY, PATRIOTIC ARMY, NATIONAL ARMY AND PROFESSIONAL ARMY: HISTORY, CHALLENGES AND THE DEVELOPMENT OF CORE IDENTITY IN THE INDONESIAN NATIONAL ARMY

Lieutenant-Colonel Eri Radityawara Hidayat, MHRMC
*Lieutenant-Colonel Gunawan, DES**

The Armed Force of the Republic of Indonesia,
Was born in a battlefield to achieve the nation's independence,
In the midst of a people's struggle to defend this independence.
Therefore, the Armed Force of the Republic of Indonesia is a:
National armed force,
People's armed force,
Patriotic armed force.

General Sudirman,
Commander-in-Chief
of the Indonesian Armed Force

Speech to the Indonesian Armed Force
and the Indonesian people,
4 October 1949[1]

Writing about the military professional identity of the Indonesian National Army or Tentara Nasional Indonesia Angkatan Darat (TNI AD) is quite a challenge. Due to its history, the TNI AD is unique amongst other armies in the world, particularly most Western armies, which consider themselves to be distinct professional entities in their societies and are regulated by certain "universal" norms.[2] However, since the aim of this book is to discuss the international perspectives of military professional identity, it is hoped that the TNI AD approach to military professionalism can be seen and understood

---

* Psychological Service of the Indonesian National Army.

Chapter 4

through its historical point of view and not only from the "traditional" perspective of military professionalism that is common to other countries.

As a direct descendant of the various militias that fought the Japanese and Dutch occupying forces during Indonesia's struggle for independence, the TNI AD was not formed by the government. In fact, it was initiated by the people who wanted to liberate themselves from colonial powers.[3] The heroism and sacrifice of the people at that time to achieve a common goal eventually resulted in the *Core Identity (Jati Diri)* of the TNI AD as People's Army *(Tentara Rakyat)*, Patriotic Army *(Tentara Pejuang)* and National Army *(Tentara Nasional)*. It was only after the enactment of a Bill of Law in 2004, however, that the TNI AD accepted the label of "Professional Army," reflecting the demand of modern society.[4]

Given that the Core Identity defines the existence of the TNI AD, its preservation is without argument, crucial. Safeguarding the Core Identity means not only inculcating the values derived from it through the TNI AD's education and training centres and the dissemination of training materials from relevant institutions, but also by assessing, selecting and developing soldiers willing to act in accordance with these values. This chapter will discuss the history and current vision of the TNI AD in relation to its Core Identity, some of the challenges to this identity, and the current effort to preserve and develop this identity through future initiatives.

PEOPLE'S ARMY, PATRIOTIC ARMY, NATIONAL ARMY AND PROFESSIONAL ARMY

In 2006, Indonesian Army Headquarters (Markas Besar Angkatan Darat – Mabesad) announced its vision as follows: "To become a TNI AD that is solid, professional, tough, nationalist and loved by the people."[5] This vision was formulated to reflect the legacy of the historical events during Indonesia's War of Independence (1945-1949) and of the modern history of the TNI AD up to the present. This eventually formed the Core Identity of the TNI AD as "people's army, patriotic army, national army and professional army" of the Unitary State of the Republic of Indonesia (*Negara Kesatuan Republik Indonesia* – NKRI).

A solid army means that every soldier must unite: they must be like brothers as well as comrades in arms. The brotherhood of the Indonesian Army soldier is based on the spiritual bond that emerges from the shared values that have been passed from one generation of soldiers to the next since the War of Independence, specifically, to always help one another at all times.[6] The solidity of the TNI AD was at its peak during the War of Independence, when virtually the entire young Republic's high ranking civilian government officials surrendered and were arrested by the Dutch colonial military ruler.[7] Under the leadership of the Commander-in-Chief (*Panglima Besar*), General Sudirman, the TNI was able to organize an underground guerrilla force that was able to reclaim the Republic's capital in Yogyakarta from the more technologically advanced and experienced Dutch forces in six hours, enabling them to prove the existence of the Republic.[8] On the other hand, the solidity of the TNI AD was at its lowest when, during the 1950s, its officer corps degenerated into factionalism and engaged in separatism that was based on political and ethno-religious groupings.[9]

The TNI AD has a unique concept of military professionalism that is somewhat different from the concept of military professionalism commonly understood in Western society. TNI AD's professionalism is not measured strictly in terms of mastering military tactics and techniques, but also based on its core identity as people's army, patriotic army and national army.[10] Therefore, a TNI AD soldier can only be considered professional if he or she can employ acquired military competencies in the context of a people's army (being one with the people, willing to dedicate his or her life for one's nation, and being part of a national army that shows no political favour). This, however, is not to offer an excuse for lacking in military capabilities, especially in consideration that professionalism in the TNI AD reached a low point when it was involved excessively in the political process of nation-building and national development, causing it to neglect the building of its military competencies.[11]

It should be noted that this situation is not unique to the TNI AD. As mentioned by Alfred Stepan in his observation of Latin American armies, the paradigm shifted from "old professionalism," of mastering purely technical military skills, into "new professionalism," in which armies in developing countries participated in the nation-building process.[12] Nevertheless, as happened with the TNI AD, Stepan noted that this approach has its drawbacks in that it can lead to the military stepping into the purely civilian domain. The TNI AD realized this fact and has shown its willingness to abandon previous involvement in the day-to-day political processes of the country, and instead, now concentrates on ways and means to improve its expertise in defending the nation.

The TNI AD soldier is also expected to be tough. The notion of "toughness" goes beyond physical endurance. It also relates to a mental toughness, akin to a fighting spirit. This means that a TNI AD soldier must be of strong character, difficult to subdue, resilient, and willing to sacrifice himself or herself for nation, country and humanity in general.[13] During the War of Independence, the Dutch forces were superior to the TNI in terms of technology, organization, personnel and other elements. Often armed with only fighting spirit and bamboo spears, the guerilla forces of the TNI were able to withstand the onslaught of the Dutch forces for four years.[14]

In modern Indonesia, where other sectors of the country require additional state funding to expand economic development, the TNI AD offers to take in less than the optimal defence budget so that the country as a whole can accelerate its development process.[15] Consequently, a TNI AD soldier must be willing to show his or her best performance regardless of the circumstances faced. This fighting spirit is the core of the TNI AD's identity as a "Patriotic Army" that is willing to defend the Republic without thought of surrender. Morris Janowitz noted that the TNI AD was one of only four armies out of more than forty that actually fought a national liberation war against a colonial force.[16] Without doubt, the TNI AD has earned the right to be considered as a patriotic *(pejuang)* army, not a professional army that is the norm in other countries.

The TNI AD's vision also states that its *soldiers are expected to have a nationalistic outlook.* The 220 million people who live in the Southeast Asian archipelago called Indonesia come from more than 13,000 islands, have descended from approximately 300 native ethnic groups and races, worship a variety of religions and speak approximately 700 distinguishable dialects.[17] Thus, the social and geographical conditions of Indonesia could serve as a recipe for disaster (as events have taught us in the former Yugoslavia, Rwanda, Chechnya, Sudan, Lebanon and other troubled spots around the world). Yet, this does not happen in Indonesia in part because the TNI AD has a nationalistic outlook that is imprinted on each and every one of its soldiers. This serves to reduce or limit the potential for social conflict. In fact, the TNI AD has consistently proven to be the national organization that boasts the greatest level of ethnic cohesion in the country.[18] In times of social conflict, a TNI AD soldier is not permitted to favour a certain ethnicity, religion, race or other grouping, including his or her own. As a reflection of the TNI AD's Core Identity, loyalty remains foremost to country and country alone.

Throughout Indonesian history, various ethnic conflicts and separatist movements have occurred from time to time. The neutrality of the TNI served as the main guarantor of the continued existence of the NKRI that is based on the national ideology of Pancasila and the 1945 Constitution. Pancasila (Five Pillars) refers to the notion that the daily life of Indonesians should be guided by five principles: belief in one God Almighty; humanity; unity of Indonesia; democracy guided by consensus; and social justice.[19] With its neutral values that do not favour one religion or ethnic group over another, time and time again, Pancasila has proven to be the unifier in times of religious and ethnic conflicts in Indonesia.[20] Similarly, the 1945 Constitution, which goes hand in hand with the declaration of Indonesia's independence, is the *raison d'être* for the existence of the Republic. It has proven to play a positive role in guaranteeing the smooth transition to more democratic periods whenever Indonesia faced various parliamentary crises, especially when the TNI leadership chose to uphold the constitution instead of using its power to achieve political ends.[21]

Lastly, *the TNI AD soldier must be loved by the people*. This means that he or she must be willing to protect the people while defending the nation and does not harm them in any way.[22] During the War of Independence, the TNI AD implemented a total defence system that relied on a territorial structure that, in turn, depended on village leaders who were trusted by the people.[23] In fact, during this time, in order to survive, the villagers took care of the well-being of the soldiers. Not only did they provide refuge to the soldiers, they also fed and clothed them.[24] It was this experience that would later influence the Core Identity of the TNI AD as the "People's Army." This means that the TNI AD is the direct descendant of the paramilitary groups that were formed by the people after the declaration of independence and then fought the colonial powers to safeguard Indonesia's independence. Thus, the TNI AD soldiers are inseparable from the people.[25]

THE EVOLUTION OF THE TNI AD'S CORE IDENTITY

The Indonesian War of Independence took place between Indonesia's declaration of independence on 17 August 1945 and the transfer of sovereignty to the Republic of Indonesia from the Netherlands on 27 December 1949. This "struggle to wrest and defend the republic's independence" *(perjuangan merebut dan mempertahankan kemerdekaan)* from the Dutch colonial power was considered one of the largest revolutions of the 20th century. It lasted over four years and involved bloody armed conflicts between the newly formed TNI and the more technologically advanced Dutch forces that had substantial combat experience from the Second World War.[26] It was from this War of Independence that the Core Identity of the TNI AD was shaped as "People's Army, Patriotic Army and National Army."

Chapter 4

History has recorded various attempts by the local people to gain independence from the Dutch who first established a permanent Indonesian presence in the early 1600s. The root of Indonesian nationalism, however, is best traced back to youth movements that grew rapidly in the first half of the 20th century to oppose Dutch colonialism. The most prominent of these was the Budi Utomo movement (1908-1935), which was led by educated Javanese (a dominant ethnic group in Indonesia), who attempted to achieve political representation in the *Volksraad* (the People's Council) on behalf of the interests of native Indonesians in the Dutch East Indies.[27] On 28 October 1928, various native youth groups declared a Youth Pledge *(Sumpah Pemuda)* at a conference that included three guiding principles: to have only one homeland, one people and one language – Indonesia. At this time, the national anthem, "Great Indonesia" ("Indonesia Raya"), was sung for the first time.[28] Considering the diversity of the people who lived in the Dutch East Indies, this pledge was indeed revolutionary. With this declaration, the concept of a united Indonesia was ignited (a successor of the Dutch East Indies) and became a rallying cry for the people.[29]

Awareness of the Indonesian identity was further strengthened when the Japanese occupying force, which came to Indonesia after defeating the Dutch forces, encouraged nationalist sentiment to further their own political goals. On 22 April 1943, the 7th Japanese Army Headquarters in Saigon proclaimed that they would provide Indonesian youth the opportunity to serve their homeland by individually joining the Japanese Army as a *Heiho* (auxiliary soldier). This proclamation was followed by an order from the Japanese military leadership in the Indonesian capital of Jakarta that instructed Indonesians to form a total defence system that would range from big cities to remote areas.[30] In effect, this system organized Indonesian settlement much like that of Japan, and in the process, introduced the "Territorial Army" concept, the Japanese warrior ethos and military discipline.

With the tides of war turning against it, it became necessary for Japan to add native auxiliary military forces in Indonesia to counter the advancing Allied forces. Therefore, on 8 September 1943, the Japanese Southern Army in Saigon issued an order to form *Kyodo Bo-ei Giyu-gun* (a voluntary army to protect the native land), resulting in the formation of the so-called PETA (Pembela Tanah Air, or Defenders of the Motherland) in Indonesia.[31] Thus, the TNI AD was born and trained by Japanese instructors in guerrilla warfare.[32] Many Indonesian youth volunteered and numbers eventually grew to 77 battalions in Java, Madura and Bali, and 55 companies in Sumatra.

Following the proclamation of independence on 17 August 1945, the Dutch attempted to re-establish their rule. Citizen militias sprung up from the various social, ethnic and religious groups, and former Indonesian volunteers in the Japanese Army became the core of the Indonesian freedom fighters. At that time, PETA officers formed the bulk of the leadership of the Indonesian Army that was established afterward to defend the young nation. Although some native Indonesians had been recruited to serve in the Dutch colonial army – the *Koninklijk Netherlands-Indische Leger* (KNIL) – very few achieved officer status.[33] In fact, as noted by Lieutenant-General (retired) Purbo Suwondo, the former commanding general of the TNI Academy, KNIL was created as an instrument of the Dutch colonial power to "counter, suppress and crush all kinds of internal rebellions."[34] He concluded that unlike PETA, which was formed specifically as territorially-based guerrillas to confront more technologically advanced invaders, the KNIL model could not become a conduit for a national freedom movement. Fortunately, under the leadership of the charismatic General Sudirman, rivalries amongst the various ethnic and religiously based pro-independence militias, and especially between the Japanese-trained ex-PETA members and the Dutch-educated Indonesian soldiers who had served in KNIL, were reconciled. This enabled the TNI to finally agree on a doctrine of "total people's resistance" instead of committing to the conventional defence method favoured by the Dutch-trained TNI soldiers.[35]

It was largely due to the Japanese legacy of creating a territorial army in order to defend Indonesia from an external aggressor with a professional armed force and technologically advanced weaponry and technical skills, coupled with the TNI AD's own experience in implementing a successful guerrilla strategy during the War of Independence, that the doctrine of "total people's resistance" was developed.[36] Within this doctrine, General Sudirman issued the order, Surat Perintah Siasat No.1, signed in November 1948 that established a system of military districts called Wehrkreise (adapted from the German system during the Second World War), in which the entire TNI would abandon linear defence and retreat to non-urban areas in order to wage guerrilla warfare.[37] Under this system of circular or regional defence, each regional commander had full authority to operate against the Dutch forces by utilizing the assets available in the district under his command.[38]

Consequently, in each military district, soldiers were required to unite with the people so that the people would be willing to supply them with the necessary logistical support, as well as provide them with information on the movement of the Dutch soldiers.[39] Recognizing the political pitfalls that might arise with this system, General Nasution, one of the greatest thinkers in the TNI AD and one of the few ex-KNIL officers who led the TNI AD, further developed this doctrine. In essence, his contributions created an Indonesian version of a guerrilla warfare doctrine that excluded the political outlook that existed in the people's army concept of most communist countries (i.e., this refined doctrine did not justify or envision the TNI AD to mirror the politically-oriented armed force that existed in those countries).[40]

After the War of Independence, from 1950 to 1957, Indonesia's civilian government was progressively weakened by traditional and ideological sentiments, so much so that approximately 100 different political parties emerged. The resulting political climate in Indonesia became so unstable that seven cabinets lasted for an average of only 15 months each.[41] Inevitably, the TNI's top leadership decided to intervene, and in 1958, General Nasution, who at that time was the Army Chief of Staff (*Kepala Staf Angkatan Darat* – KASAD), proposed the concept of *Dwi Fungsi* (Dual Function), in which the military would assume a middle of the road position, meaning that it would be neither a political player like experienced in Latin American armed forces, nor would it become a spectator like the Western European armed forces.[42]

When parliament accepted this concept and it became law, the TNI then had a legitimate socio-political function in addition to its defence and security function. With the failed "communist coup" of 1965, which eliminated most of the senior leadership of the TNI AD, General Suharto then created an *"Orde Baru"* (New Order) regime and he rose to power to become Indonesia's president for the next three decades.[43] The territorial command structure had proven its success in forcing the Dutch to capitulate and recognize Indonesia's independence. Unfortunately, the dual function concept slowly evolved and deviated from its original meaning. It later became a justification for Suharto to stay in power with the support of the army's territorial structure down to the village level.[44] With this derailment of TNI AD's unique identity, its bond with the people at this time, especially with the more educated and politically active urban society, was broken.

When President Suharto's government collapsed following the Asian financial crisis in 1998, the *Dwi Fungsi* doctrine fell apart.[45] With the birth of the "Reformation Era," the TNI formulated a new approach called *Paradigma Baru* (New Paradigm) as part of an internal armed force reform. Under *Paradigma Baru*, the TNI willingly agreed to dismantle its socio-political role and instead concentrate on its role and function as a defender of the nation. This marked a return to the Core Identity.[46] With this "New Paradigm," TNI AD soldiers were expected to adopt new attitudes and conduct themselves more in accordance with a modern professional army, while retaining the Core Identity as "People's Army, Patriotic Army and National Army." One of the results of this "New Paradigm" was the addition of the "Professional Army" component to the Core Identity, which was legalized in 2004 through the Bill of Law no. 34.[47] This change was praised by Western experts who appeared more comfortable with a notion of professionalism that emphasized added focus on efforts to improve war-making capabilities, especially against external enemies.[48]

CHALLENGES TO THE TNI AD'S CORE IDENTITY

The advent of globalization has posed a significant challenge to the glue (i.e., the pluralistic ideology of *Pancasila* and the 1945 Constitution) that has held the Indonesian nation together for more than six decades. Some pundits observed that the decline in the central government's authority as a result of political decentralization that took effect after the financial crisis and the resulting downfall of President Suharto's rule have increased rivalries and created ethnic and religious conflicts that threatened the very existence of the NKRI.[49] In short, globalization pressured people to conform to a uniformity of meaning. People became fearful of the erosion of their local cultures and this fear created a longing for a connection with a cherished past. Eventually, this fuelled religious fundamentalism and ethno-nationalism which, in the past two decades, has spread all over the world, across all ethnic groups and religions.[50]

Indonesia, of course, is not immune to this phenomenon and surveys have shown that more and more Indonesians are conforming to their religious duties.[51] Many trends have suggested that as a Muslim majority country, Indonesia is becoming "more Islamic" and radicals have tried to portray Indonesia as a haven for terrorists, bent on turning the country into an Islamic theocracy. The most recent comprehensive survey on religion and identity in Indonesia demonstrated, however, that a solid majority of Indonesian Muslims prefer a democratic state. Considering its 100-year tradition of tolerance, the emerging picture confirms that Islamic radicalism cannot be equated with mainstream Islam in Indonesia.[52]

The problem with religious fundamentalism in a pluralistic society like Indonesia occurs when those aspiring to create a society based on their strict values attempt to import "global values" from a more homogenous society in another country and then try to change the social contract that was enshrined in the national ideology of *Pancasila* and the 1945 Constitution. Although they constitute a small minority, the political openness that came about during the Reformation Era created an opportunity for those who espoused fundamentalist aspirations that were banned during the New Order. In what the religious scholar Marty Martin and historian R. Scott Appleby called "delayed psychological response," these groups made no secret of their long held and repressed ambition of turning the country into a religious state, and in the process, erode the country's identity and ideology.[53]

In his landmark book *The Professional Soldier*, Morris Janowitz showed that the values of a military organization in any country will always reflect the dominant values of the society that it serves.[54] The TNI AD's organizational culture and identity have been shaped through a long history. Nevertheless, values that were laid down by the founding fathers of the TNI AD have now been challenged because of the spread of new global values in Indonesian society and by the changing events in contemporary Indonesia.[55] Although the TNI AD with its Core Identity has become a model of multiculturalism in Indonesia, it is still inevitable that in an era of globalization, some of its soldiers will want to promote a radical aspiration that is contrary to the Core Identity that stresses that one must put the nation's interest above the interest of a particular region, ethnicity, race or religion.

Another significant challenge to the TNI AD's Core Identity is the infighting and disputes among civilian politicians that produce unworkable governments and an unstable political climate. Forty-eight parties contested the 1999 general election after more than 150 political parties registered at the Ministry of Justice and Human Rights (many failed to fulfill the election administrative requirements). This political excitement continued in the 2004 general election.[56] In these elections, ideologies ranged from the socialist-leaning to the religious. In reality, this kind of situation is not unique to Indonesia. For example, the transition from a single communist party to a pluralistic democracy in Poland resulted in a mushrooming of no fewer than 111 political parties that contested the 1991 Polish parliamentary election.[57] As experience demonstrated during the 1950s, the political manoeuvres of civilian politicians ultimately dragged the TNI into the political arena because they felt that it was their obligation to defend the interests of the nation as a whole. As Samuel Finer observed in his study of military interventions in a nation's political process, countries that have weak political institutions coupled with constant power struggles among politicians are more likely to see their armed force intervene.[58]

Fortunately, unlike many other developing countries, the Indonesian military did not enter into politics by way of *coup d'état,* but in an orderly and legal fashion.[59] As General Wiranto, the Commander-in-Chief of the TNI during the waning days of the New Order-era in 1998 stated, it would have been very easy for the military to stage a coup at that time, but they did not do so out of respect for the 1945 Constitution that precludes a military government.[60] In fact, during the years of military dominance in the New Order, never once did the TNI seize power. As John Hasseman, an expert on the Indonesian military asserted, "Indonesia can be said to have a government with a powerful military, but not a military government."[61] The issue has been more centred on civilian politicians and the government of the day who were tempted at times to seek support from the military to further their own political agenda.[62] This was the case during the latter days of the New Order-era, so much so that the military was forced to become involved in petty politics and support the ruling political party.[63]

Unfortunately, this is also happening in contemporary Indonesian politics, in which immature civilian politicians still attempt to seek support from members of the TNI by provoking them to get involved in the political process, even as the military establishment sends a strong signal that they are unwilling to do so.[64] Based on the Indonesian experience, in the implausible event that the military should decide to again involve itself in the day-to-day political process, it will surely lose the support of the people. And if this were to happen, the bond between the soldier and the people will with certainty be weakened and the Core Identity of the TNI AD as an army of the people will gradually fade.

PRESERVATION AND DEVELOPMENT OF THE TNI AD'S CORE IDENTITY

The TNI AD utilizes several methods to preserve and develop its Core Identity as "people's army, patriotic army, national army and professional army." The most important of these is the educational and training system, as well as the efforts of the Mental Guidance Service of the Army *(Dinas Pembinaan Mental Angkatan Darat* – DISBINTALAD). Naturally, TNI AD's Education and Training System (*Komando Pendidikan dan Latihan Angkatan Darat* – KODIKLAT) is responsible to produce graduates who will live by the TNI AD's Core Identity.[65] The KODIKLAT is responsible for the TNI AD's educational curriculum. It has a directorate for education and a directorate for doctrine, and is responsible for the entire curriculum development process for each educational level of the TNI AD officer, non-commissioned officer and soldier.[66] This centralized system was created to ensure that curriculum development is designed systematically, sequentially, seamlessly and in an integrated fashion to fulfill the overall educational objective of the TNI AD. Apart from KODIKLAT, the Indonesian Staff and Command School (*Sekolah Staf dan Komando Angkatan Darat* – SESKOAD), as one of the central operational units reporting directly to the *Kepala Staf Angkatan Darat* (Army

Chapter 4

Chief of Staff – KASAD), has the responsibility of delivering the highest level general development course and conducting strategic analysis for the TNI AD.[67] As a result, SESKOAD has a very important role in formulating the strategy to preserve and maintain TNI AD's Core Identity, as well as in creating future TNI AD leaders who will adhere to those values.

Value inculcation at SESKOAD and KODIKLAT is hinged on three basic tenets of the TNI AD training and educational system (*Tri Pola Dasar*): academic achievement, physical fitness and character building.[68] Character building refers to the process of how to create a soldier who will be aware of, accept, and then demonstrate the Core Identity. Each training and education centre has a department solely devoted to this endeavour, called the Fighter Department (*Departemen Kejuangan* – DEPJUANG), which basically is responsible for delivering mateials that can enhance the students' partiotism. In the past, materials developed and taught by DEPJUANG were more akin to indoctrination, covering among other subjects the history of the TNI AD, the state ideology of Pancasila and the TNI AD's Core Identity.[69] As an example, at SESKOAD, students receive traditional classical classroom courses on *Kejuangan* subject matter supplemented by a week long *Kejuangan* seminar called *Program Kegiatan Bersama (PKB) Kejuangan* (Joint Staff and Command Activities to instill Kejuangan values). This includes students from the SESKOAD, SESKOAL (Navy Staff and Command School), SESKOAU (Air Force Staff and Command School) and foreign students who listen to prominent members of the TNI, including the commander-in-chief, the service chiefs and former freedom fighters, deliver lectures and war stories.[70] The students are then graded through examinations, assignments, participation in seminars and observations by lecturers and syndicate leaders.

Since 2006, however, and in line with the "New Paradigm," SESKOAD has attempted to implement a competency-based curriculum in which knowledge transfer and TNI AD values are expected to be acquired through open dialogue, critical reasoning and discourse. This represents a considerable departure from the indoctrination method of the past.[71] The current curriculum is designed to stimulate the ability to think in an integrated way. Whenever possible, the doctrinal method of learning with programmed "school solutions" is minimized. Instead, lecturers encourage students to consider the possibility of different ideas coming from class discussion. An example of significant change involves the assessment of the student's ability. SESKOAD abandoned the use of the conventional examination method and now requires students to write essays. This enables the assessor to better appreciate the depth of the students' thoughts on the subject, including knowledge of *Kejuangan* materials. Syndicate discussions, moderated by Facilitating Officers (*Perwira Penuntun* – PATUN), are also used to discuss a student's work and progress. The facilitating officers are responsible to stimulate students thoughts and assess their ability to solve problems. This technique is especially useful when *Kejuangan* materials on modern times are being promoted.

In order to anticipate the rapid changes that occur in the world, SESKOAD also attempts to socialize students to the new TNI AD organizational culture. This is accomplished by instilling the understanding of strategic management and managing change, including leadership, as critical elements that shape the culture of the organization. Students also receive instruction on team building and group dynamics.[72] In general, the current educational approach used at SESKOAD focuses on pushing students to think in an integrated way and to be able to solve problems comprehensively and to do so under intense pressure. In addition, SESKOAD demands that students become leaders who can mentor future leaders, build teams, think in a critical and creative way and improve the organization they lead.

The assessment method at SESKOAD is based on the concept of competency-based learning: students are assessed on their ability to learn subject materials independently, including the ability to conduct in-depth learning, and to develop and apply the subject materials in a large forum. Basically, these abilities are assessed by the PATUN. To help ensure objectivity and consistency in the assessment process, the PATUN remains with the same syndicate throughout the course.

Apart from the educational and training system, the Mental Guidance Service of the Army also serves to instill the values and identity of the army. In fact, since the reformation, and in line with the "New Paradigm," the DISBIN-TALAD has been responsible for ensuring that soldiers return to their Core Identity as "People's Army, Patriotic Army, National Army."[73] Apart from providing religious services much like the chaplain corps in other armed forces around the world, the DISBINTALAD is responsible for maintaining the *Kejuangan* tradition.[74] DISBINTALAD is also responsible to develop and maintain the *Kejuangan* tradition to support and maintain the Core Identity of the TNI AD (i.e., instilling fighting spirit, the willingness to sacrifice and heroism). In addition, the DISBINTALAD publishes books, pamphlets and audiovisual materials, operates army museums and preserves TNI AD history. A unique product of the DISBINTALAD is the *Bintal Fungsi Komando* (Command Function of Mental Guidance), essentially a management method to assist TNI AD commanding officers to guide their soldiers to embrace the Core Identity.

Chapter 4

Another method to maintain the Core Identity of the TNI AD is through legislation. One of the problems facing the TNI as an institution is the absence of sufficient regulations concerning national defence and security issues.[75] Since the reformation-era, however, the government and the Indonesian Parliament have formulated bills of laws on this matter. As mentioned, a tenet of the TNI as an institution is the Core Identity (i.e., defined as "People's Army, Patriotic Army, National Army and Professional Army.")[76] This legislation provides the legal foundation to preserve the Core Identity of the TNI AD and helps deter attempts to deviate from it.

AN ARMY OF THE FUTURE

Indonesia is a country endowed with richness unparalleled anywhere in the world. It has a strategic location, fertile land, an abundance of raw materials, and a long history of tolerance and moderation amongst its multi-ethnic and multi-religious society. Yet at the same time, this richness has commanded considerable international attention. Its spices were fought over by colonial powers for hundreds of years and its diversity has recently become a source of division that threatens the nation's stability. Fortunately, the idea of a united Indonesia runs very deep in the Indonesian consciousness. Through the TNI, Indonesia has remained united since the War of Independence. People from all ethnic and religious groups participated in the struggle for independence, both against the Japanese and the Dutch colonial powers.[77]

For all of its perceived weaknesses, the TNI AD is the only institution in Indonesia with a national network that reaches down to the local village level.[78] Furthermore, despite its excesses in the New Order-era, independent surveys have shown that the majority of Indonesians still see the TNI AD's territorial command structure (*Komando Territorial* – KOTER) as vital to the maintenance of law and order and public safety.[79] With its Core Identity as "People's Army, Patriotic Army, National Army and Professional Army" it has oftentimes become a source of stability in a sea of chaos. It is therefore imperative that the TNI AD of the future carry on its professional duty to defend the country while maintaining its time-proven Core Identity. The question then becomes, how does one create a modern professional fighting force, while at the same time retaining the Core Identity?

The radical transformation from Cold War confrontation to information age defence preparedness and warfare means that war will be fought and won not merely by physical force, but through the capabilities of highly competent military personnel.[80] Consequently, military forces have been compelled to improve the way they select, train and manage the performance of competent soldiers. In fact, many armies are adopting a competency-based system and using it for selecting and developing officers, performance management and instilling organizational values.[81] In this regard, the TNI AD is no exception. Since 2004, the Psychological Service of the Indonesian Army (*Dinas Psikologi Angkatan Darat* – DISPSIAD) has utilized an assessment-centre method to select candidates for the positions of the main territorial (resort and district military commanders) and combat commands (battalion commanders). The assessment-centre competency framework consists of behavioural, technical and organizational competencies that were derived from studies of civilian and military competency frameworks from other organizations, as well as DISPSIAD's own research on the criteria for success in these positions.[82]

Two methods were used to develop this framework: values-based and research-based competency analysis. Values-based competencies, which were considered unchangeable and should reflect the original TNI AD Core Identity, were derived from TNI AD values and attributes.[83] In contrast, research-based competencies were derived from the behaviour of successful performers (commanders) and could be modified in accordance with the changing environment.[84] Therefore, these competencies should reflect the addition of a "Professional Army" component to the TNI AD identity.

While this program, which is called the Position Competencies Assessment Program (PCAP), is still in its early stage of development, DISPSIAD has proposed that it be made part of a comprehensive Competencies-Based Human Resources Management System (CBHRM) that could be used for selection, development and performance management.[85] As an example, through collaboration between DISPSIAD and Personnel Staff of the Army (*Staf Personel Angkatan Darat* – SPERSAD), the behavioural competencies can be used as a basis to design performance appraisal tools and leadership development programs. On the other hand, the KODIKLAT and SESKOAD can develop the technical competencies to ensure that future commanders have the required professional capabilities to conduct warfare at various levels. The organizational competencies cluster, which consisted of TNI AD values that reflect the original Core Identity, can be used by DISBINTALAD to design development programs that will ensure TNI AD's soldiers will remain close to the people, put the nation's interest above anything else, and maintain an unyielding fighting spirit.

Thus far, this initiative has shown promising results. In terms of assessment and selection, the competency framework has provided a clear focus on what to look for and is considered fair by the candidates.[86] Currently, the TNI AD is implementing a pilot project on a leadership development program for junior officers based on the competency framework at the company commander level. Although the jury is still out on the success of this project, it is hoped that if the CBHRM system is implemented, the TNI AD will develop the leadership necessary to put into practice the Indonesian Defence White Paper vision (i.e., being professional, not engaged in politics, subordinate to the democratic government, educated and well trained, equipped with the required weaponry and compensated accordingly).[87]

CONCLUSION

The TNI AD, which is the largest standing army in Southeast Asia and the South Pacific region, has a proud tradition as a liberation army that was born in the midst of a battlefield during Indonesia's War of Independence. It is because of this experience that the TNI AD did not become a "traditional" professional military institution as commonly understood in the Western world. Instead, the TNI AD formed its own Core Identity as "People's Army, Patriotic Army and National Army." Considering the diversity of Indonesia as a developing country, the TNI AD with its Core Identity has shown to be one of the few institutions that can hold the country together and preserve the social contract, namely the *Pancasila* and the 1945 Constitution (which makes Indonesia a singular entity). The TNI AD has also shown its flexibility in adapting to the changing times as evidenced by its willingness to add the "Professional Army" component to its identity.

As long as the TNI AD can preserve its identity, it will continue to exist and perform its duties in defence of Indonesia. History has also taught us, however, that the powers of the TNI AD can be abused, that societal values can change, and its Core Identity can be threatened. When this happens, if the TNI AD is unable to reflect on its reason for being, then its Core Identity will fade and its role will be questioned. It is therefore critical for TNI AD officers to always think creatively on how to preserve and develop the Core Identity so that the TNI AD as an institution can exist for centuries to come.

> We should create TNI soldier who is aware of his/her Core Identity,
> A soldier who does not know his/her identity does not have the
> moral strength to defend the interest of the nation.
>
> > General Endriartono Sutarto, Commander-in-Chief of TNI,
> > Briefing to the Joint Staff and Command School students,
> > 4 August 2005.[88]

ENDNOTES

1. Pusat Sejarah TNI, *Soedirman & Sudirman* (Jakarta: Pusat Sejarah TNI, 2004), 49. This quotation was taken from a speech given by General Sudirman, the Commander-in-Chief of the Indonesian Armed Force (TNI) in 1949 during the height of Indonesia's War of Independence.

2. Mabesad, *Setia dan menepati janji serta Sumpah Prajurit* [Loyal, Persistent and Committed to the Soldier's Oath] (Jakarta: Markas Besar Angkatan Darat, 2006), 16-17.

3. Barry Turner, "Nasution: Total People's Resistance and Organicist Thinking in Indonesia" (Unpublished PhD Thesis, Swinburne University of Technology, Melbourne, 2005), 70.

4. Letkol Inf Imam Santosa, *TNI sudah berusia 62 Tahun, lalu bagaimana?* [TNI is Already 62 Years Old, Now What?] (May 2008), http://www.tni.mil.id/news.

5. Mabesad, 2006, op.cit., 10.

6. Ibid.

7. Ibid., 13.

8. Yahya Muhaimin, *Perkembangan militer dalam politik di Indonesia, 1945-1966* [Military Development in Indonesian Politics, 1945-1966] (Yogyakarta: Gadjah Mada University Press, 1982).

9. Mabesad, *Sejarah perjuangan kepemimpinan TNI Angkatan Darat* [The History of the Struggle of the Indonesian Army Leadership] (Jakarta: Markas Besar Tentara Nasional Indonesia Angkatan Darat, 2005), 89-96.

10. Mabesad, 2006, op. cit., 10.

11. Mabesad, 2006, op.cit., 13-14.

12. Alfred Stepan, "The New Professionalism of Internal Warfare and Military Role Expansion," in Alfred Stephan, ed., *Authoritarian Brazil* (New Haven: Yale University Press, 1973).

13. Mabesad, 2006, op. cit., 18.

14. Sadao Oba, "My Recollections of Java during the Pacific War and Merdeka," *Indonesia and the Malay World* 8, no. 21 (1980), 6-14.

15. Mabesad, 2006, op. cit., 13-14.

16. Morris Janowitz, *The Military in the Political Development of New Nations* (Princeton, NJ: Princeton University Press, 1964), 1-53.

17. Adrian Vickers, *A History of Modern Indonesia* (Cambridge: Cambridge University Press, 2005).

18. William Zhow, "Indonesia: Military Reform and Modernization," *Military Technology* 24, no. 12 (2000), 36-41.

19. Eka Darmaputera, *Pancasila and the Search for Identity and Modernity in Indonesian Society: A Cultural and Ethical Analysis* (Leiden: Brill Academic Publishers, 1997).

20. Leo Suryadinata, "Nation-Building and Nation-Destroying: The Challenge of Globalization in Indonesia," in Leo Suryadinata, ed., *Nationalism and Globalization: East and West* (Singapore: Institute of Southeast Asian Studies, 1999), 38-70.

21. William Liddle, "Indonesia's Democratic Transition: Playing By the Rules," in Andrew Reynolds, ed., *The Architecture of Democracy* (Oxford: Oxford University Press, 2002), 373-399.

Chapter 4

22. Mabesad, 2006, op. cit., 29-31.

23. Abdul Nasution, *Fundamentals of Guerrilla Warfare* (New York, NY: Praeger, 1965), 54, 112.

24. Tahi Simatupang, *Laporan Dari Banaran* [Report from Banaran] (Jakarta: Pembangunan, 1961).

25. Indonesian Army Headquarters, *Profile of the Indonesian Army* (Jakarta: Mabesad, 2007), 33.

26. Theodore Friend, *Indonesian Destinies* (Cambridge, MA: Harvard University Press, 2003), 35.

27. Amry Vandenbosch, "Nationalism in Netherlands East India," *Pacific Affairs* 4, no. 12 (1931), 1051-1069.

28. James Sneddon, *The Indonesian Language: Its History and Role in Modern Society* (Sydney: UNSW Press, 2003).

29. Jusaf van der Kroef, "The Term Indonesia: Its Origin and Usage," *Journal of the American Oriental Society* 71, no. 3 (1951), 166-171.

30. Joyce Lebra, *Japanese-trained Armies in South East Asia: Independence and Volunteer Forces in World War II* (New York: Columbia University Press, 1977).

31. Nugroho Notosusanto, *The Peta Army during the Japanese Occupation of Indonesia* (Tokyo: Waseda University, 1979).

32. Angel Rabasa and John Haseman, *The Military and Democracy in Indonesia: Challenges, Politics, and Power* (Santa Monica: RAND, 2002), 8.

33. Bilveer Singh, "Civil-military Relations in Democratizing Indonesia: Change amidst Continuity," *Armed Forces & Society* 26, no. 4 (2000), 607-633.

34. Purbo Suwondo, "The Genesis of the Indonesian National Army and some Political Implications," Paper presented at the *International Seminar of the Institute of Netherlands History and The Royal Society of Historians of the Netherlands*, 27-29 March 1996, The Hague, Netherlands.

35. Anonymous, "Personality Profile: General Sudirman," *The Pointers, Journal of the Singapore Armed Forces* 33, no. 3 (2007), http://www.mindef.gov.sg/imindef/publications/pointer/journals.

36. Kelompok Kerja SAB, *Sedjarah Singkat Perdjuangan Bersenjata Bangsa Indonesia* [Short History of the Armed Struggle of the Indonesian People] (Jakarta: Staf Angkatan Bersendjata, 1964), 74.

37. Petra Groen, *Marsroutes en dwaalsporen - Het Nederlandse militair-strategische beleid in Indonesië 1945-1950* [Lines of March and Wrong Tracks: Dutch Military Strategic Policy in the Dutch East Indies 1945–1950] (Den Haag: SDU, 1991).

38. Nugroho Notosusanto, *Pejuang dan Prajurit, Konsepsi dan Implementasi Dwifungsi* ABRI (Jakarta: Sinar Harapan, 1984), 55.

39. Himawan Soetanto, *Yogyakarta 19 Desember 1948: Jenderal Spoor (Operatie Kraai) versus Jenderal Sudirman (Perintah Siasat No. 1)* (Jakarta: Gramedia Pustaka Utama, 2006), 269.

40. Robert Cribb, "Military Strategy in the Indonesian Revolution: Nasution's Concept of Total People's War in Theory and Practice," *War and Society* 19, no. 2 (2001), 144-154.

41. Mabes ABRI, *Pengantar Sishankamrata* [Introduction to the People's Security and Defence System – Sishankamrata] (Bandung: Sekolah Staf dan Komando ABRI, 1993).

42. Salim Said, *Legitimizing Military Rule: Indonesian Armed Forces Ideology, 1958-2000* (Jakarta: Pustaka Sinar Harapan, 2006).

43. Robert Elson, *Suharto: A Political Biography* (Cambridge: Cambridge University Press, 2001), 186-187.

44. Anonymous, "…and, of course, order: Indonesia is bringing its army under discipline. Now it needs a serious police force," *The Economist*, 6 July 2000, http://mail.tku.edu.tw/113922/Economist_Survey.

45. Jenderal TNI, Endriarto Sutarto, *Kewajiban prajurit mengabdi kepada bangsa* [The Obligation of the Soldier is to Serve the Nation] (Jakarta: Pusat Penerangan TNI, 2005), 5-7.

46. Mabes TNI, *Bunga Rampai Paradigma Baru TNI* [Compilation of TNI's New Paragigm] (Jakarta: Staff Komunikasi Sosial, Mabes TNI, 2003).

47. Santosa, 2008, op. cit.

48. John Bradford, *The Indonesian Military as a Professional Organization: Criteria and Ramifications for Reform* (Singapore: Institute of Defence and Strategic Studies, 2005), 19.

49. Alan Tidwell and Charles Lerche, "Globalization and Conflict Resolution", *International Journal of Peace Studies*, 9, no. 1, (2004), 47- 59.

50. Shmuel Eisenstadt, "The Resurgence of Religious Movements in Processes of Globalisation – Beyond End of History or Clash of Civilizations," *International Journal on Multicultural Societies* 2, no. 1 (2000), 4-15.

51. Anonymous, "Makin Salah Makin Curiga," *Tempo* [More Mistakes, More Suspicions] (December 2001), 29. The *Tempo* weekly cited a survey conducted by Pusat Pengkajian Islam dan Masyarakat (PPIM), Universitas Islam Negeri, Jakarta.

52. Jamhari, "Radical Islam and the Consolidation of Democracy in Indonesia," in Myra Torralba, ed., *Acting Asian: Contradictions in a Globalizing World* (Tokyo: International House of Japan, 2004), 43-62.

53. Martin Marty and R. Scott Appleby, eds., *Fundamentalisms Observed* (Chicago: The University of Chicago Press, 1991), 9.

54. Morris Janowitz, *The Professional Soldier: A Social and Political Portrait* (New York: The Free Press, 1971).

55. Mayor Jenderal Tippe, MyarifudinSi, "Strategi pengembangan TNI AD 25 tahun ke depan: Ditinjau dari perspektif pendidikan" [TNI AD's Development Strategy for the Next 25 Years: The Educational Perspective] *Yudhagama* 70 (2006), 14-28.

56. Aris Ananta, EviArifin and Lurvidya Suryadinata, *Emerging Democracy in Indonesia* (Singapore: Institute of Southeast Asian Studies, 2005), 4-5.

57. Kenneth Ka-Lok Chan, "Poland at the Crossroads: The 1993 General Election," *Europe-Asia Studies* 47, no.1 (1995), 123-145.

58. Samuel Finer, *The Man on Horseback: The Role of the Military in Politics* (New Brunswick: Transaction Publishers, 2003), 86-89.

59. Salim Said, *Legitimizing Military Rule: Indonesian Armed Forces Ideology* (Jakarta: Pustaka Sinar Harapan, 2006).

60. Damien Kingsbury, *Power Politics and the Indonesian Military* (London: Routledge-Curzon, 2003), 163.

61. John Hasseman, "To change a military – the Indonesian Experience," *Joint Force Quarterly* 29 (2000), 23-30.

Chapter 4

62. Marcus Mietzner, *The Politics of Military Reform in Post-Suharto Indonesia: Elite Conflict, Nationalism, and Institutional Resistance* (Washington: East-West Center, 2006), 16.

63. Leo Suryadinata, "The Decline of the Hegemonic Party System in Indonesia: Golkar after the Fall of Soeharto," *Contemporary Southeast Asia* 29, no. 2 (2007), 333-358.

64. Anonymous, "Yudhoyono Ingatkan Politikus Sipil Tak Goda TNI/Polri" [Yudhoyono Remind Civilian Politicians Not to Tempt] *Sinar Harapan*, 26 January 2004.

65. Sayidiman Suryohadiprojo, *Si Vis Pacem Para Bellum: Membangun pertahanan negara yang modern dan efektif* [If You Seek Peace, Prepare for War: Building Modern and Effective National Defence] (Jakarta: Gramedia Pustaka Utama, 2005), 237-247.

66. "Kodiklat", Retrieved 5 May 2008 from http://www.tniad.mil.id/kotama/kodiklat.php.

67. "Seskoad", Retrieved 5 May 2008 from http://www.tniad.mil.id/lemdik/seskoad.php.

68. "Seskoad", *Petunjuk pelaksanaan tentang penilaian hasil belajar Perwira Siswa Dikreg Seskoad* [Operational Directive on the Evaluation of Study Results of Indonesian Army's Staff and Command School Students] nomor Skep/7/I/2006 tanggal 27 January 2006, 2.

69. Bradford, 2005, op. cit.

70. Seskoad, *Jam Kadep Juang Seskoad kepada Pasis Dikreg XLIV Seskoad TA 2006*, Lecture by the Head of the Kejuangan Departmen for the 2006 Seskoad students.

71. Seskoad, *Reforming Seskoad's Curriculum*, Presentation prepared for the Subject Matter Expert Exchange between SESKOAD and the US Army Command and General Staff College, 5 June 2005. As SESKOAD is the most important career course for the Army, and even though the discussion in this paper only covers the changes at SESKOAD, it can be assumed that the rest of the army's educational system will also be patterned after it.

72. Tim Orasi Ilmiah Seskoad, *Merancang kurikulum berbasis kompetensi di Seskoad* [Designing Competency-Based Curriculum at SESKOAD] Presentation made at the closing ceremony of the 2005 SESKOAD, 27 October 2005.

73. Ety Juminatun, *Pembinaan mental tradisi kejuangan dalam upaya peningkatan ketahanan mental prajurit TNI AD dan implikasinya bagi ketahanan bidang Hankam di wilayah kota Yogyakarta (Studi di Korem 072/Pamungkas)* [Mental Guidance for the Fighter Tradition to Enhance the Mental Resilience of TNI AD Soldiers and its Implications for Defence and Security in Yogyakarta Area (A Study of Korem 072/Pamungkas)] (Unpublished Master's Thesis, Gajah Mada University, Yogyakarta, 2004).

74. "Disbintalad", Retrieved 30 April 2008 from http://www_tniad_mil_id.htm.

75. Andi Widjajanto, "Evolusi Doktrin Pertahanan Indonesia" [Evolution of Indonesian Defence Doctrine] in Ikrar Bhakti et al., eds., *Kaji Ulang Pertahanan: Perspekif Politik* [Defence Review: Political Perspectives] (Jakarta: LIPI, 2005).

76. "Undang-undang Republik Indonesia Nomor 34, tahun 2004 tentang tentara nasional Indonesia, Bab II, Jati Diri" [Bill of Law of the Republic of Indonesia Number 34 Year of 2004 on the Indonesian Armed Force, Chapter II on Core Identity], retrieved 4 May 2008 from http://www.tni.mil.id/.

77. Benedict Anderson, "The Future of Indonesia," in Michel Seymour, ed., *The Fate of the Nation-state* (Montreal: McGill-Queen's University Press, 2004), 386-387.

78. Patrick Walters, "The Indonesian Armed Forces in the Post-Soeharto Era," in Geoff Forrester, ed., *Post-Soeharto Indonesia. Renewal or Chaos?* (Singapore: Institute of Southeast Asian Studies, 1999), 59-60.

79. Lembaga Survei Indonesia, *Publik Merasa Kekuatan Teritorial TNI Masih Penting* [The Public felt that the TNI's Territorial Structure Commands are Still Important], Survey conducted in May 2006, retrieved 4 May 2008 from http://www.lsi.or.id/riset/104/sikap-publik-terhadap-institusional-tni.

80. Ryan Henry and C. Peartree, "Military Theory and Information Warfare," in Ryan Henry & C. Peartree, eds., *The Information Revolution and National Security* (Washington: CSIS Press, 1998).

81. Eri Hidayat, "A Case Study of the Use of a Competency Framework in the Australian Army for Performance Management and Development" (Unpublished Master's of Human Resources Management and Coaching research thesis, University of Sydney, Australia, 2005).

82. Lieutenant Colonel Gunawan and Eri Hidayat, "The Use of Assessment Center in the Indonesian Army," Paper presented at the *19th International Military Testing Association (IMTA) Conference*, Gold Coast, Australia, 10 October 2007, retrieved 4 May 2008 from http://www.imta.info/PastConferences/ PowerPoints.aspx.

83. William Steele and Robert Walters, "21st Century Leadership Competencies: Three Yards in a Cloud of Dust or the Forward Pass?" *Army Magazine* 51, no. 8 (2001), 31.

84. US Army Combined Arms Center, *The Army Training and Leader Development Panel Officer Study Report to the Army* (Fort Leavenworth: CAC, 2001).

85. Eddy Koesma and Eri Hidayat, "Assessment Centre for Performance Management in the Indonesian Army," Paper presented at the *2nd Indonesian Assessment Centre Congress*, Jakarta, Indonesia, 26 July 2007.

86. Gunawan & Hidayat, 2007, op.cit.

87. Dephan, "Buku Putih Dephan" [Indonesian Department of Defence White Paper], retrieved 4 May 2008 from http://www.dephan.go.id/buku_putih/.

88. Sutarto, 2005, op.cit., 119.

CHAPTER 5

PROFESSIONAL IDEOLOGY
IN THE NEW ZEALAND DEFENCE FORCE

Squadron Leader Murray Simons

> Three stonemasons were building a cathedral when a stranger wandered by. The first stonemason was toting rocks to a pile, near a wall. "What are you doing?" asked the stranger.
> "Can't you see that I'm carrying rocks?"
> The stranger asked the second labourer, "What are you doing?"
> "I'm building a wall to exact dimensions," he replied.
> A few steps away, the stranger came upon a third mason. "What are you doing?" he asked. This worker smiled. "I'm building a cathedral!"

Arguably, all three stonemasons belong to a profession, two are being professional, but only one embraces a professional ideology.[1] Could this parable have a modern military equivalent? How many soldiers see themselves as just doing a job? How many believe they are professional in *applying violence*,[2] and how many also have some concept of a higher purpose? Does it make a difference as to how we work when we know the greater goal of who we are and to what we are contributing? Should our professional ideology be completely aimless and responsive only to external and uncontrolled forces? If indeed we want to maximise our effectiveness, then surely we must harness this powerful phenomenon. But before stewarding a professional ideology, it is necessary to not only live by it, but also to have a deep understanding of its tenets.

This chapter will explore the subject of professional ideology from a New Zealand perspective, more specifically, from a New Zealander's perspective. While every effort was made to canvas widely on the subject, much of this is opinion and based on anecdotal evidence. At the time of writing, very little domestic debate had occurred and almost no empirical data existed. Furthermore, this is a dynamic and evolving subject of increasing interest to many military professionals. It is anticipated that the New Zealand Defence Force's (NZDF) collective understanding of the subject will continue to mature in time. Therefore, much of what is written here remains, at best,

Chapter 5

relevant only at the time of writing. Of course, the intent of this book is to promote interest in the subject and invite debate.

This chapter opens with a brief exploration of professional ideology as a concept before exploring the NZDF's perspective. The first section begins with the national influences before considering pan-NZDF and single Service dynamics. Overall, this chapter places greater emphasis on the cultural dimension of ideology; treating the specialist knowledge aspect as relatively robust, both in content material and delivery system. The abstract nature of the military culture makes it more complex and therefore consumes more attention. A clear distinction is made between the official *espoused* culture and the reality of the normative *in-use* culture. Attention is also afforded to the delta, or difference, between these two concepts of the NZDF's culture. Discussion then centres on recognising and controlling this delta, primarily by influencing the in-use culture. This chapter concludes with the assertion that the greatest weakness in the NZDF's professional ideology is the poor appreciation many members have of their contribution to the organization's goal.

PROFESSIONAL IDEOLOGY –
A NEW ZEALANDER'S PERSPECTIVE

Professional ideology, as a term, is only just entering NZDF vocabulary. It is neither in common use around the camps and bases, nor – with the exception of the newly adopted Professional Military Development (PMD) Framework – does it appear in the written products of higher headquarters. This would, therefore, make for a very short chapter if it were not for the fact that the phenomenon exists, just without a label. Thus, this book contributes to a relatively new but exciting journey of exploration into a much needed, yet poorly understood, academic debate in the NZDF.

Just as militaries have slightly different understandings of courage, duty and honour, it would appear also that there are different understandings of terms like ethos, ideology and culture. Because some esoteric terms have multiple, similar, and at times, even conflicting interpretations, it is important to be clear about the nuances intended in this chapter. Except where indicated, all terms and concepts are drawn from Dr. Bill Bentley's seminal work *Professional Ideology and the Profession of Arms in Canada*.[3]

Ideology is a multidisciplinary concept used to articulate a complex understanding of an institution's *raison d'être* and cultural identity. It captures both the ethos and specialized knowledge unique to that society. Embedded within this complex mix is a clear understanding of the ultimate goal for which the organization strives. This goal should not be confused with the poetic vision statements that adorn letterheads and business cards. Rather, it is a clear understanding of the organization's higher purpose. In the case of the military, this will include clear linkages between national (government) goals and directed operational activities.[4] But this is perhaps the easiest part of understanding an ideology.

Like culture, an ideology defines its members as much as they define it. While a formally espoused ideology might be preserved in documentation, the in-use version will be dynamic. Daniel Bell goes so far as to suggest ideologies "not only reflect or justify an underlying reality, but once launched…take on a life of [their] own."[5] Figure 1 is an attempt to visually represent the main interdependencies of this amorphous phenomenon in the military context.

Bentley claims ideologies are systematic, normative and pragmatic.[6] By this he means they have a structured approach to how their sub-topic knowledge and beliefs interrelate. Their understanding is guided, and corrected, by the collective members – not imposed by higher authority. And they give guidance to enacting tangible behaviours or actions. In short, they provide the members with both the organization's goals (ends) and *modus operandi* (ways). While this breakdown is useful, clarification is necessary on the normative aspect. An organization can have two ideologies – one is their official (espoused) version and the other is the actual (in-use) ideology. While these two should be the same, the living (normative) version will be potentially more dynamic and constantly vacillating. Ideologies exist in many disciplines – some of which include religious, political, economic, social, and even various professions.

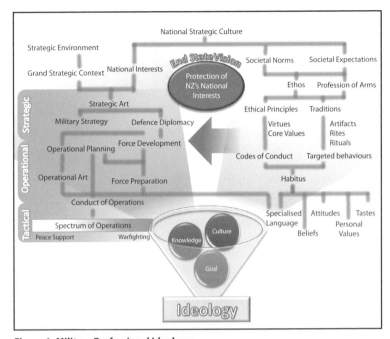

Figure 1: Military Professional Ideology.

Professional ideology captures the essence of a profession. It combines the ideals (ultimate goal) with both the specialist knowledge and ethos of a given profession. While different professions will have their own respective purpose and areas of specialist knowledge, their ethos will likely share a number of common elements. It is this commonality between professions that distinguishes them from other business ideologies. Eliot Freidson, for example, clearly distinguishes professional ideology from market and bureaucratic ideologies. Where market ideology is based on consumerism, competition and profits, bureaucratic ideology seeks structure and standardization.[7] In contrast, professional ideology is characterised by the tenets of a profession, in our case, the Profession of Arms.

The Profession of Arms shares many of the attributes of other professions but with a military flavour. While no singularly agreed upon definition exists, many authors have explored the topic in the past 50 years.[8] Once again, this chapter defers to Bentley's useful synopsis of the leading commentators.[9] Table 1 outlines the NZDF's key attributes of the Profession of Arms.

The Profession of Arms

- Provides a higher purpose to society

- Has a monopoly on specialized skills

- Is self-regulating

- Has its own ethical value system

- Is responsible for its own development

- Has a robust sense of community

- Has lawful allegiance to government

- Has society's trust & confidence

Table 1: The Profession of Arms from the NZDF's Perspective.

The Profession of Arms Ideology differs significantly from Market and Bureaucratic Ideologies. As has been frequently highlighted, the scale and legacy design of conventional military forces means they are, by default, also bureaucratic. Furthermore, because most Western militaries recruit from a market-oriented society, they import self-interest and an internally competitive ethos into their ranks.

Unfortunately, because the three-way relationship is zero-sum,[10] the acculturation of market ideology and need for bureaucracy both continuously challenge the primacy of professional ideology within the military. While the three-circle *Competing Ideologies* model depicted by Bentley[11] implies discrete categories, a more dynamic construct is proposed, *with profession symbolically moved to the top* (see Figure 2).

As Charles Moskos discovered, individuals seldom remain static in their cultural and work ethic identity.[12] Although his earliest versions of the institutionalism-occupationalism (I/O) construct[13] depicted discrete boxes, he soon revised the thesis into a linear spectrum. He determined that individuals might move freely along the continuum depending on postings, operational tours and other personal influences, while an organization is typically more stable. Short of strategic shock events such as the impact 9/11 had on the US military, most organizations will move fairly slowly depending on the size (relative number and impact of sub-groups) and stability of the culture (based on strength and depth of traditions, rituals and artifacts). Ideology however, is on a different level to Moskos' I/O construct.

Chapter 5

As shown in Figure 2, the three sides of the *Competing Ideologies Triangle* are continua along which individual members – or an organization's collective norm – can be plotted. Because the forces act in three directions, a person (or organization) can appear anywhere in the triangle. At the top of the triangle, professional ideology aligns with Moskos' concept of institutionalism, while the bottom – and especially market ideology – correlates (but not synonymously) with occupationalism, or self-interest. Bureaucratic ideology aligns, to some extent, with the principles of management, while professional ideology emphasizes the art of leadership. A case could also be made that market ideology has parallels with the concept of command. The notion of *profit above all else* would translate to *mission success above all else*. As with the Command, Leadership, Management (CLM) trinity, all three ideologies have a place in the military, but the right balance is important.

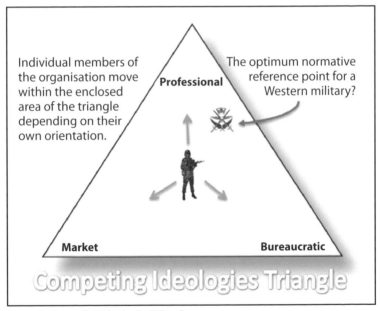

Figure 2: Competing Ideologies Triangle.

The three competing ideologies found within the military ebb and flow in their influence. Because ideologies are normative, they reflect the dominant aspects of the collective group. When the majority of members are disproportionately closer to one particular apex, the organization's defining ideology will shift. With a predominance of self-interested (occupationalism-orientated) members, the organization would appear closer to the market apex. Because the normative culture can itself also be an influencing factor in member orientation, the phenomenon is partially self-stabilising. Institutional inertia dampens cultural change.

Bureaucratic ideology, for all its faults, is also a necessary evil. As previously noted, the massive scale of most military operations demands that bureaucratic systems manage both materiel and personnel. Therefore, the optimum position for a conventional military would be close to professional ideology but not exclusively. An extreme of any ideology is considered unhealthy.

A purely professional ideology is not ideal for any military. Near-extremist professional ideologies (and their associated indoctrination techniques) appear in many guises, from the mountain caves of Afghanistan to the boot camp of Parris Island. These organizations risk increasing the delta between themselves and their society's ethos. But the militaries of democratic nations need to reflect and protect the ideals of the people they serve – not control them. At the absolute extreme, militaries whose ethoi diverge significantly from their host society lead *coups d'état*, while in less severe cases, culturally isolated militaries risk losing the confidence of their society (a tenet of a profession). Examples of this were apparent with the Canadian Forces in Somalia, US Forces in Abu Ghraib and the New Zealand (NZ) Police with recent group-sex scandals coupled with perceived systemic protectionism. Even when the delta is small, however, there can still be an impact on both recruiting and retention.

As the pendulum of society swings, so too must the military's. While the nature of professional ideology, and the need to breed internally, means the military will typically remain right of centre (more conservative), they cannot nail their normative foot to the floor (fundamentalism). As discussed earlier, the military culture needs to reflect its society. When generational waves (Gen X, Y, and millennials) move through the ranks, they shift the normative reference point and the military must accommodate their needs. The challenge though, is not to move too fast. While remaining relevant and attractive to future generations is important for recruitment and retention, fully embracing market ideologies devalues both the military covenant and the Profession of Arms.

Understanding the dynamics of the competing ideologies triangle is fundamental if we are to preserve the Profession of Arms. While mercenary armies and private security forces might work for some countries, New Zealand in particular cannot afford to adopt a market ideology for its Defence Force. Unlimited liability and concepts of patriotism are critical to success in the highest end of conflict – survival of the nation. Success in other areas is similarly reliant on collectivist values.

Those who devise quirky recruiting incentives to attract and retain Generation Y members need to consider the destructive legacy their initiatives might leave. Allowing civilian consultants to determine what attracts recruits is only half the equation. The military does not just need new recruits, it needs the right recruits. A high attrition rate during initial training is indicative of an untenable delta between their personal ideology and the military's. Perhaps a greater concern is not those who leave, but those who remain and begin negatively influencing the organization's culture. Weakening the military's professional ideology reduces combat effectiveness.

But it is not just the recruiters who can have a strategic impact on the organization's culture. Those in higher headquarters who continue to impose bureaucratic and civilian financial models on operational-level commanders need to realize the damage they are also doing to combat effectiveness. Just as market ideologies erode the vestiges of our profession, so does excessive bureaucracy. Innovation, passion and creativity are virtues that permit inspirational leadership and success on operations. Somewhere, a balance must be found.

PROFESSIONAL IDEOLOGY IN NEW ZEALAND

> New Zealand has a proud military heritage that has contributed to the cultural make-up of our nation. The military has served with distinction in past conflicts, due in part to the make-up of our citizen forces. New Zealand's society is a liberal, secular and tolerant one that exhibits a unique 'Kiwi' combination of cultural characteristics and values. Our people have the reputation of being egalitarian, friendly, of good humour, honest, individualistic, open-minded and creative.
>
> The NZDF Joint Vision Statement.[14]

NZDF personnel are both products of and remain members of the greater New Zealand culture. Most aspects of the society therefore permeate across, however, some do not. While the military reflects society, it traditionally remains "right of centre." Despite the so-called Generation X and Y (and now millennials) debate, new recruits and officer candidates are typically drawn from either military families or those with collectivist ideals. This means that those who survive initial recruit training and excel in the meritocracy system are usually those who exemplify the desired culture. This self-selection system helps perpetuate the traditional military culture with relatively stable direction. It also preserves the cultural alignment with society.

New Zealand's strategic culture is heavily influenced by its short history. The first known inhabitants arrived around 700 years ago. Within a few hundred years, the first Europeans began visiting, and by the 1800s, the country was developing as a multicultural nation. Over the next 200 years, various waves of immigrants created a steady population increase. Most of the immigrants were from Europe, but other significant groups were from Asia and the Pacific. Typically they were in search of better lives and were escaping the tyranny of poverty and of class systems. They wove their social and political ideologies into the rich tapestry that continues to shape our pluralist nation today.

The settlers of the past 200 years were typically tough individuals who could endure harsh conditions and had the drive and vision to create better lives. The few landed gentry who purchased large land holdings were quickly disempowered by legislation. Popular action forced the dismantling of large estates and deliberately abolished any chance of class differentiation. This concept of egalitarianism remains strong today and is often cited as an identifying feature of New Zealand's service personnel by other forces.

The survival attitude of the early settlers also gave rise to another typical Kiwi characteristic – improvisation. From the beginning, settlers in the new land learned to adapt and modify whatever they had to help build their better lives. During hard times, such as the World Wars, New Zealanders enhanced this characteristic by becoming a nation of hoarders. Garden sheds and kitchen cupboards filled with recyclable spare parts. Known colloquially as the "Number eight fencing wire" mentality (after the popular gauge of farm fencing wire used to fix just about anything), Kiwis are proud of their inventiveness and creative problem-solving techniques. This national trait remains strong within the NZDF and leads to another popular concept known as the "can do attitude." A reluctance to say "no," however, has found the NZDF over-stretching itself on many occasions.

Equally influential in New Zealand's identity is the indigenous culture. While the waves of immigration created many issues, Maori culture endured and is now enjoying substantial resurgence across all sectors of New Zealand society. The acculturation of Maori tikanga (customs and traditions) is particularly noticeable in the Defence Force. As a warrior culture, Maori people excel in the military and their ethnic ethos is not only complementary, it is embraced.

Chapter 5

In New Zealand, public support for war has typically been consistent with other Western nations. Small numbers volunteered for the Boer War, but much larger contributions were made to the two World Wars. Korea and Vietnam had smaller contributions and once again reflected the growing unease publics around the world had for fighting abroad during the 1960s and 1970s. Both anti-war and anti-nuclear sentiments of this era saw commensurate declines in public support for the military as a whole.

During the twilight years of the Cold War and into the so-called peace dividend era of the New World Order, the military was seen as a comfortable vehicle for gaining qualifications. Parents and grandparents encouraged their charges to join the military and get a head start in life. This saw the rise of market ideologies in the military that, at the time, was identifying a decline in institutionalism. The flurry of interest by academic and senior military leaders in Moskos' I/O construct highlighted the concern of this era. But really, this was just a reflection of the changing society.

During the 1980s and 1990s, New Zealand, along with many other Western nations, was experiencing a rise in individualism. This attitude was driven mainly by the secondary impact of economic discourse. Neoliberal reforms shaped a generation of self-centred individuals who were products of an education system that promoted *human capital theory* over *social capital theory*. As Codd identifies, "Within the ideology of neoliberalism, education is viewed as a product and schools are seen as being similar to small business firms."[15]

The commodification of education not only treated students as economic outputs, it bred a generation of individualists who equated success with accumulating credits. Governments assumed increased individual earning power would eventually translate into collective economic success for the country. But, the fiscally driven school curricula were only part of a grander stewarding of the national culture. Selling-off state-owned enterprises and the restructuring of remaining government departments saw a rapid change in cultural direction. What were once pure professions were now being forced to adopt market ideologies. Health, education and many other government departments were expected to transmogrify into business models. The important issue for the militaries (who largely survived the reforms) was not the shift to market ideologies by their sister professions, but the impact on future workforces due to the commodification of education. But this radical experimentation may well have been short lived.

In what was termed "the Third Way," the governments of New Zealand, Australia and the UK (and probably others) claimed to move back to the "old style socialism of the Keynesian welfare state" around the turn of the millennium.[16]   While many critics argue the return has not been enough, the past ten years have seen some resurgence of social capital theory promoting democratic citizenship. It is also true that full-fee paying international students still appear in most schools and the national qualifications system is still based on bite-sized credits (market ideology) rather than holistic subject coverage (professional ideology). Members of the "*what's in it for me*" generation should not be blamed for their selfishness; they are arguably, at least to some extent, products of government policies that shaped society's culture.

It is too early to tell whether this prophesized, reverse pendulum swing will translate into greater social capital and a commensurate reduction in market ideology within the military. Research in the United States, however, suggests that things are indeed beginning to change. Numerous commentators claim the emerging *millennial generation* is returning to the collectivist ideals of the Baby Boomers. Citing a number of these studies, Art Fritzon *et al.* assert, "Like their grandparents, millennials appear deeply committed to family, community, and teamwork, which they have made priorities."[17] They make a number of other encouraging conclusions in the area of trust, jointery (i.e., two or more services working together), innovation and decision-making. The fruits of the new educational approach and the *millennial generation* are only just arriving on recruit courses and it will be another decade before they are the dominant workers in the Profession of Arms.

The military of today has a well-known high operational tempo and the idea of joining the Defence Force for a free education is over. Those enlisting post-9/11 appreciate the reality of "unlimited liability" and the "military covenant." Compared with earlier cohorts, they are acutely aware that they could be sent off to some inhospitable corner of the world and put in harm's way. Although there are no recent empirical data to support the thesis, it is likely those who enlisted in the past five years or so have a stronger concept of professional ideology than those of the 1990s.[18]

PROFESSIONAL IDEOLOGY IN THE NZDF

As already discussed, professional ideology involves not only appreciating a higher-purpose but also a complex mix of both specialist knowledge (in our case, military operations) and culture. To evaluate the current state of professional ideology in the NZDF, it would be necessary to test the collective understanding of our purpose, the level of expertise in our specialist knowledge area, and the strength of the culture. The latter could in turn be considered by either the espoused or in-use culture. The extent and depth of all four aspects is difficult to define, let alone measure.

A simplistic answer to the state of professional ideology in the NZDF is that it is good. Based on anecdotal evidence and feedback from combined operations, New Zealanders hold their own. Competency in specialist knowledge appears on par with like-minded militaries, and indeed, the development systems are not only globally benchmarked, but are intertwined. With no domestic Professional Military Education (PME) at the strategic level, all formal courses are international. This external reliance exposes a gap in specialist knowledge and an absence of study in the unique setting of New Zealand's geo-political environment. This is only one of many shortfalls in the NZDF's professional ideology.

The ethos aspect of professional ideology in the NZDF is far from faultless. In the absence of a significant "failure of command," it is difficult to know if there is a ticking cultural time bomb waiting to go off. Foreign military examples suggest such issues usually emerge in sub-groups (individual units) under certain conditions. But waiting for the slices of Swiss cheese to align is a reactive approach to identifying an unacceptable delta. A better way would be to proactively evaluate the in-use cultural climate and conduct a subjective gap analysis with the espoused cultural identity. To do this, it is first necessary to review the espoused culture.

> The NZDF will measure the quality of its performance in a number of ways, including formal benchmarking, lessons learned and coalition feedback. The NZDF will identify the highest levels of performance considered achievable considering its size and skills, and then achieve them through a process of continuous improvement throughout the organization.
>
> The NZDF Joint Vision Statement.[19]

THE NZDF's ESPOUSED CULTURE

The espoused culture of the New Zealand Defence Force appears in a number of unrelated documents. These span the three Services as well as some for the pan-NZDF and Defence civilians. They range in focus from vision statements, definitions of respective ethoi, governing principles, codes of conduct, targeted behaviour posters and core value statements. The remaining part of this section summarizes the key tenets of these documents across the NZDF.

At the macro level, the NZDF is pragmatic enough to acknowledge the fluidity of the national culture, but reinforces the institutional inertia that dampens the impact of fads:

> Although the characteristics of New Zealand society will gradually evolve over time, the NZDF's fundamental values will continue to determine how it fulfills its responsibilities to the nation and its citizens. The NZDF is a product of the characteristics of the New Zealand's people, and as an organization it will continue to appreciate and nurture these characteristics as part of its military values and ethos.[20]

The vision statement also acknowledges the contribution of self-selection in the maintenance of this professional ideology:

> The NZDF's personnel will exhibit strong self-discipline, augmented by emotional maturity along with high levels of political and cultural understanding. The NZDF will strive to recruit, grow, care for and retain the right people so that the NZDF is as well prepared as it can be. It will need people who already practice the values of the NZDF and who can be trained and educated to a high level of expertise in their chosen field.[21]

In addition to the joint vision, the NZDF's espoused ethos appears well developed and comprehensive. Despite a number of perhaps overly confident imperatives, the *Joint Vision Statement* offers a straightforward explanation of what the ethos is, and at a macro level at least, how it is to be maintained. The following two paragraphs provide a useful window into the current state of macro-level thinking on the NZDF's espoused ethos:

> The NZDF's ethos is a warfighting ethos. It is a living spirit, describing the enduring values, beliefs, expectations and professional standards of the organization and creating and shaping NZDF culture. This ethos is embraced by all service personnel and also by all NZDF civilian staff, whose vital work enables the nation's warfighters. The NZDF's ethos will continue to unite the organization and keep it strong, providing the vital moral, physical and emotional capacities to maintain and sustain the initiative in both peace and war. At its core are the strong and enduring common values that reflect New Zealand and its history. The close and continuous two-way connectivity between New Zealand's society and military ensures that this relationship will remain inextricably linked. Those who join the NZDF are shaped by the existing ethos; they also contribute to its evolution.

Chapter 5

The NZDF's ethos consists of five interdependent uniting components. The NZDF will continue to be bound by the vital purpose that originates from the unique burden of responsibility it carries. The NZDF will maintain strong communities of trust across and within New Zealand and the NZDF, being comprised of New Zealand as well as maintained for New Zealand. The NZDF will maintain its focus on leadership that promotes discipline, judgement, innovation and confidence. The NZDF will continue to adhere to the principles of the New Zealand Profession of Arms that acknowledges the NZDF to be a distinct self-regulating organization with a monopoly on a particular expertise, performing a vital and higher purpose in the interests of New Zealand and its people. The strong and enduring core values of the NZDF will remain.[22]

At the exo, meso and micro levels, other sub-cultures exist within the NZDF. Most levels have even published their own espoused culture. At the exo level, all three single Services have produced numerous communiqués to articulate their ethoi, culture, core values and governing principles. Many have gone on to produce resources for local commanders and regularly promote their culture through in-house magazine stories and posters. The diversity and extent of them is in itself a positive sign.

Of the three Services, the NZ Army's culture appears the most examined. In the past two years, a small team has done extensive work in identifying and promulgating their ethos, values and target behaviours. This development work has arguably set an example for the other Services in terms of both depth and breadth. Posters, articles and resources can be found in every workplace, but they are not without their critics.

Some people remain cynical about the value of formally articulating something that should be dynamic and reflective. Another criticism is the lack of alignment with the pan-NZDF culture. For example, nowhere does *Joint Effect* appear in their key documents. They frequently talk about working with other nations and coalitions but not the other NZDF Services.

The Royal New Zealand Navy (RNZN) has also spent considerable effort in defining their culture. Their three core values – *Courage, Comradeship* and *Commitment* – appear on nearly every official document and are well known to all members. The *Three Cs*, as they are known, is a catchy alliteration capturing the essence of this small nation navy. The bumper sticker slogan however, is noticeably different from the NZ Army's (and the NZDF's) core values of C3I (*Courage, Comradeship, Commitment and Integrity*).

The RNZN's culture is not without other more serious critics. While most naval members loyally subscribe to the organization's direction, those outside it watch the *civilianization* with scepticism. Their extensive use of civilian management models places greater tension on the bureaucratic-professional ideology spectrum at the operational and tactical levels. To be fair, this drive has backed off slightly and is no longer pushed at all levels. Furthermore, the general direction has endured several chiefs and may now be systemic. The other two services however, have clearly not followed.

The Royal New Zealand Air Force (RNZAF) also has a well-published cultural identity in terms of ethos, mission and values. Interestingly though, these all differ significantly from those of the other two Services. When choosing their espoused ethos, the RNZAF considered adopting the same as the NZ Army (warrior), but discounted it because it sounded too land-centric – too army. The decision was instead made to select "warfighting" over "warrior." Semantics aside however, the concept is effectively similar. Perhaps the alternate word choice says more about the RNZAF's desire to stand apart from the army than it does about who or what they identify with. As it transpires, warfighting ethos is now the NZDF's official term too. As with the independence shown in the ethos description, the RNZAF also differs significantly with its values.

While the army and navy share almost identical values to the stated NZDF values, the RNZAF has retained its unique list. Furthermore, the current chief of air force remains adamant that there will be no alignment or subjugation of the NZDF values simply to show unity. While this may be interpreted as another act of defiance and independence from the other Services, it is also a reflection of integrity. As the chief rightly points out, the current list of values is a contract. The list was generated by a cross-section of RNZAF personnel who were chosen to represent the collective organization. Conversely, those "other values" appear to have been conjured up in some closed-door conference room without consulting the people who they are supposed to represent.[23]

THE NZDF's IN-USE CULTURE

The NZDF's in-use culture is different from its espoused one. While impossible to explore all aspects, the dimension of jointery is indicative of both the differences and efforts to reduce the delta. As indicated above, the respective single Service culture documents reveal rifts in cultural unity, despite this being a stated organizational goal.

Chapter 5

Ever since the dark era of strategic level, and very public, inter-Service rivalry, the NZDF has sought a stronger joint culture. At the turn of the century, fiscal constraints saw a campaign by senior army officers against the RNZAF's air combat force.[24] While the resultant loss of this NZDF capability remains today, the subsequent fallout, when their tactics became known, was a complete cleanout of "the old guard." New service chiefs were found and a replacement chief of defence force (CDF) was plucked from two ranks below. His first job was to reduce the delta between the government's espoused expectations and the in-use culture at the time. But the need to reduce the inter-service rivalry delta was not unique to New Zealand.

The NZDF followed the global trend of increased jointery by standing up more and more pan-organization units. Cautious to learn from the Canadian experience, the NZDF did not fully embrace a unified system, but rather looked for appropriate areas for synergistic gains. One of the first was a Joint Force Headquarters to control all operational activities. Removing this important aspect from the Service Chiefs left them free to focus on the raise, train and sustain role. The subsequent explosion of joint, integrated and pan-NZDF pre-modifiers heralded a new direction in the organization's culture.

Stewarding the culture is a complex and slow process. Of the many artifacts – or levers – available to the (then) new CDF, was the NZDF's by-line. With a strong publicity campaign behind it, the catch phrase "Three Services, One Force" was intended to emphasize the strength in unity behind the three strands of the same rope. Over time, the slogan morphed to, "Three Services as One Force". Its current version reads "Three Services as One Force, being the best in everything we do." Each iteration progressively attempts to reduce the macro-level delta between the espoused culture of jointery and the in-use inter-Service rivalry. But cultural deltas transcend every layer of the organization.

At the exo level, all three Services have ongoing problems with vices. By definition, these are also values of a collective group, only the opposite extreme of virtues. But drinking alcohol and aggressive competitiveness are not likely to appear on any culture posters. Isolated pockets within military communities indicate that there are other even less palatable cultural identifiers. From court-martial findings and newspaper articles, it is apparent the NZDF has groups of people who use recreational drugs, commit domestic violence and steal from their comrades. To be fair, these small-group sub-cultures are not condoned by the organization and are not representative of the wider NZDF community. Their extreme behaviour, however, does effect shifting the norm reference of the collective away from the espoused culture. They also indicate the diversity of members within the collective group.

THE NZDF's CURRENT FOCUS

The NZDF's current focus should be on developing all facets of professional ideology. This includes: improving the level of understanding and commitment to our purpose, increasing specialist knowledge, and managing the organization's cultural direction – vice ensuring the right espoused culture is determined and widely promulgated; monitoring the in-use culture; and leading changes to maintain an acceptable delta. This latter task is perhaps the most challenging. Because culture is not only abstract and slow to change, many initiatives need to be part of "the long game." Strategic vision is necessary to anticipate both problems and solutions. Stewards need to be skilled in addressing not only complex, but wicked, problems.[25] And while many great leaders have natural skills in this area, all benefit from additional education.

Maintaining and enhancing current specialist knowledge has become far more important in the current climate. Significant and continuous changes in the international security environment mean that the Cold War approach of dispersed episodic courses is no longer effective. Military personnel need to embrace a culture of continuous lifelong learning. Furthermore, because of changes in technology, broader education in the management of violence (vice simply training in its application) needs to be given to lower rank levels. The rise of the strategic corporal[26] phenomenon now demands a wider focus – and greater conceptual understanding – when promoting professional ideology.

Stewarding the profession and its ideology first requires an understanding of what these are. Coupled with the need to understand the various components of this sociological phenomenon, practitioners need to be skilled at employing their interrelationships and shaping the cultural battlefield. Ironically, those who are entrusted with this very important task have usually only been groomed in warfighting tactics, not applied sociology. But things are changing. Increasing attention is now being paid to educating the higher echelons about the theory of ideology and its associated tenets. *Stewarding* now appears on senior non-commissioned members' (NCM) courses and should soon also appear on officers' PMD programs. But understanding the theory of ideology is only the first step.

The tenets of the military's professional ideology permeate all PMD curricula and are increasingly appearing in syllabi. With greater awareness of the subject, what was previously only ad hoc and personality-driven is progressively being documented as part of a course's non-formal learning outcomes. Through the use of a matrix system, declarative knowledge syllabus objectives can now be cross-referenced to other learning activities from the *parallel curriculum* to en-sure structured professional ideology development (see Figure 3). The visual reference table shows at a glance the interdependencies of learning outcomes and the relative emphasis being placed on key tacit learning. What was previously neglected by the throwaway label of *intangible benefits* is increasingly being documented in course curricula. Now, course designers and deliverers can clearly see why and how formal and non-formal learning activities have been chosen the way they have. Additionally, non-formal and informal activities can be better appreciated for their contribution to holistic learning. Sequencing and relative value is also more apparent. While not a panacea, this quasi-subjective approach to the abstract art of enculturation is a start.

Target Quality	Formal						Non-Formal					In-Formal				
Cultural Intelligence Enhancement	Syllabus Objectives						Curriculum Activities					Encouraged Activities				
	CLM 1.3	CLM 4.2	STRAT 1.7	IR 8.2	IR 8.3	OPS 3.2	Team Presentations	Mid-Course Cocktail party	Internal Day	Overseas Study tour	Embassy visits	TV Lounge	Shared Fitness	Morning Teas	Officers Mess	Peer review of essays
Meta culture knowledge	●		●					●		●		●	●	●	●	●
Cultural Awareness		●					●	●	●	●	●	●	●	●	●	●
Cultural Tolerance			●	●			●			●		●			●	●
Cross-cultural learning		●					●			●					●	●

Figure 3: Sample Matrix showing Cross-referencing of Learning Activities.

Education is as complex as training is complicated. While training can have multiple moving parts that interact with each other, it is often simply a hierarchical taxonomy of procedural learning. Education on the other hand is complex. This means it is more spontaneous, unpredictable, irreducible, contextual and vibrantly sufficient.[27] The need for holistic and synergistic packages is vital in developing the right character in service personnel. The focus of development, however, varies with the level of training. Junior courses are more likely to indoctrinate the core military values, while higher-level programs focus on the enculturation of more strategic orientated belief systems. Regardless of the focus though, all PMD needs to be clear in both its intent and emphasis on intangible learning outcomes.

Most training units exploit opportunities to shape the organization's culture by moulding students' habitus (a set of acquired patterns of thought, behaviour and taste in an individual). While some of those on the receiving end may simply "play the game" by *accommodating* the desired behaviour, others will internalize the values. Factors shown to influence this difference include: developmental age of recipient, extent of previously established conflicting values, duration of exposure and intensity of methods used.[28] Entry and junior level military courses typically *indoctrinate* trainees through inculcation of expected values while later courses employ more subtle techniques. Using well-crafted formal learning activities, expected behaviour is deliberately *encultured* while non-formal learning (due to the environment and atmosphere) encourages *acculturation* through peer group cross-pollination. Space precludes a thorough examination of these processes here, but further reading on the following concepts is recommended: assimilation, conditioning, socialization, infusion, habitus, proselytizing, the hidden curriculum, the parallel curriculum, heuristic learning, tacit learning, formal, non-formal, informal and incidental learning, Kolb's experiential learning, Kohlberg's stages of moral reasoning, Krathwohl's Taxonomy of affective domain learning, Kegan's Identity Development, the Ashen Model, and social learning theory – to name but a few.

Developing the right professional ideology involves both reducing negative influences and promoting positive ones. Having taught the middle/senior leadership about the theory of professional ideology, and offering course designers a mechanism to include the process in course curricula, it is also necessary to identify what exactly should be enculturated. Despite the constant evolving nature of our *specialist knowledge*,[29] the updating and delivery of this *professional expertise* aspect is managed within a mature teaching/training system. The content of the ethos aspect, however, is not always so clear. The espoused culture is perhaps a useful starting point.

The NZDF's Joint Vision Statement should be the basis for cultural development on all Joint courses. As an overarching document that addresses culture, ethos, values and belief systems, this is a useful and authoritative source for the writing and rewriting of all Defence College curricula. Similar single service documents (which should align with the NZDF one) should likewise be used for their respective courses.

Espoused culture statements do not capture everything about the ideal NZDF. For example, the threat of competing ideologies does not yet feature in any NZDF cultural document. Both workplace policy writers and course designers/deliverers must remain cognizant of the subversive impact market and bureaucratic ideologies can have. As previously discussed, a degree of both is necessary in a democratic nation's military, provided they are not at the expense of professional ideology as the normative reference. But because formal PMD courses are so influential in cultural reproduction and cultural change, they need to minimise market and bureaucratic forces. They also need to be structured in how they progressively develop all facets of professional ideology.

The NZDF has adopted the Canadian Forces' Professional Development Framework (PDF) for aligning the numerous single service and pan-NZDF courses. Space precludes a revisit of this concept here but several publications expand on the concept.[30] It is, however, worth re-emphasising that this framework posits Professional Ideology as central to all aspects of professional development. The NZDF intends to complement their version of the Framework with a series of publications to help designers and deliverers of PMD enhance cultural alignment.

There are many ways to steward a professional ideology. Long before sociological concepts were ever taught in the military, effective leaders were successfully shaping culture. Today though, with the benefit of labels, models and theories, this innate ability can be enhanced even further. Legacy systems still permit reactive behaviour modification for breeches of acceptable conduct, in particular, military law. But the more subtle art of proactive enculturation is now becoming equally deliberate.

Both cultural change and cultural reproduction are, and should be, consciously targeted by commanders of all ranks. At the macro level, strategic leaders develop artifacts (such as vision statements, ethos paragraphs and values posters) as well as policies to help align the organization's in-use culture with their espoused one. These not only create pan-organizational changes but also provide fertile environments – complete with the tools – for local commanders to implement unit level stewarding. But this alone is not enough.

CHALLENGES

Stewarding a professional ideology is indeed challenging. In an attempt to clarify the diverse and overlapping issues, the main challenges have been broken into five categories (see Figure 4). These include promoting the organization's purpose, improving specialist knowledge, managing both the espoused and in-use cultures, and reducing the resulting cultural delta. While many of these have been touched on under the Current Focus section, others deserve further consideration. This final section therefore, will review the major challenges.

Specialist knowledge is perhaps the easiest component of professional ideology to maintain, but it is not without challenges. As is well known, the rapid changes to the global security environment mean that doctrine, and even principles, are being constantly rewritten. While some may say change has always existed, the need for the modern warrior to remain academically current remains vital. But academic study must compete with operational deployments and workplace demands in the zero-sum fight for time. Current personnel shortages in the NZDF, coupled with a higher operational tempo, places a strain on releasing personnel for courses and on course lengths – shorter courses compromise content coverage.

Recent advances in learning theory are also challenging the way Defence Colleges teach. Emerging generations of military students are demanding more modern approaches to their learning. Old-school lectures and passive *instructivism* need to be replaced with interactive and engaging opportunities where students can maximize their learning within *constructivist* learning environments.[31] Changing to this style can be a challenge for military staff that are more comfortable in the dichotomous paradigms of tradition training.

The NZDF has never placed much emphasis on teaching sociology. While inter-personal psychology is often included on leadership development courses, the broader concept of collective social dynamics – such as stewarding a profession – has been largely ignored. Leaders at all levels in the NZDF need to receive professional education in why, what and how culture can be observed, measured and stewarded.

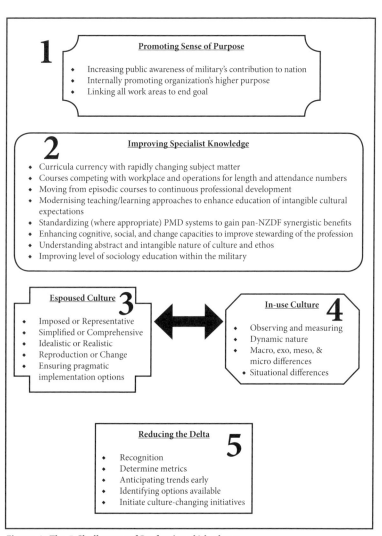

Figure 4: The 5 Challenges of Professional Ideology.

Although easier than the other aspects, determining an espoused culture also has many challenges. Some of which include who should write it – senior leadership or a cross-sectional representation of the organization – and how simple should it be? To leave it overly simplistic opens gaps in interpretation while overly complex makes it incomprehensible. Then there is the question of how idealistic and inspirational it should be, versus how accurately it should reflect the actual in-use culture. This ultimately questions whether it seeks cultural change or cultural reproduction. And then there is the issue of how pragmatic should the espoused culture be? For decades, core-values posters were hung on walls simply "to ward off evil spirits." Without being linked to daily activities, they are meaningless. Training courses were often equally at fault. With nicely worded attitudinal goals listed in the curriculum, there was seldom a link or mechanism to translate them into real learning. Fortunately, this is being partially addressed with instruments such as the matrix shown in Figure 3.

However, the greatest challenges by far involve managing the in-use culture. In particular, the perennial problems that once saw the demise of the trait approach to leadership apply equally to culture. While the behavioural-era idea of listing then teaching the traits of great leaders was a worthy ideal in its day, the rise of cognitivism, and later humanism, highlighted many challenges. As was discovered, measuring, comparing, describing and teaching intangible personality characteristics is problematic. Consequently, such reductionist approaches drifted out of popularity for many years.

Of note, the emergence of sophisticated technology and related research in the areas of both psychometric testing and learning theory has seen a revival of the trait approach. Attempts to develop personnel based on personality traits are now considered worth revisiting. However, as Spencer and Spencer identify, "core motive and trait competencies at the base of the personality iceberg are more difficult [than surface knowledge] to assess and develop; it is more cost effective to select for these characteristics."[32] They go on to criticize organizations that recruit based on university educated employees in the false-hope of reaping the right trait and motivation. Fortunately, the NZDF select their mainstream officers based more on potential (personality, ability and motivation) than academic ability – and so perpetuate the organization's ethos, culture and professional ideology. But this does not negate the need for proactive development.

How do you measure comradeship or commitment? We often only notice breaches of values when we need to impose discipline. Thus shaping values is typically reactionary and in terms of negative response vice proactive. Medallic recognition and other intrinsic reward systems such as meritocratic promotion, however, are examples where positive displays of values are formally recognized. While these levers are instruments of both cultural reproduction and change, it is the former that faces daily threats from emerging generations and external forces.

Chapter 5

Newer members of the NZDF bring their societal values with them. But they alone are not the only threat – in fact, self-selection means they are often more military than the military.[33] Misguided bureaucrats, both uniformed and civilian, in higher headquarters can be equally damaging. In their quest to achieve short-term goals (for example, recruiting or retention quotas), those with great influence on the culture often overlook the longer-term impact of their new initiatives. For example, replacing service-housing communities with financial offsets promotes extrinsic motivators at the expense of professional ideology tenets.

Bureaucratic ideologies are revealed in management principles and are essential for the efficient and effective organization of a large institution. Despite this, management is often despised by lower and mid-level leaders because it constrains the archetypal attributes of professional ideology. In the zero-sum time tension between leadership and management, most military leaders want to exploit flexibility and innovation to achieve their mission and look after their people. Ironically, without the bureaucratic shackles of procedures, a military full of mavericks would quickly degenerate into anarchy. The need for militaries to reflect society (including a degree of market ideology) has already been discussed. The dynamic between the three competing ideologies therefore needs managing, but is not considered a significant problem. Competing professions, however, could be a different issue.

The most obvious military personnel who identify with multiple professions are classified as specialists, which in itself reinforces their differentness from *real* Profession of Arms members. These include doctors, dentists, chaplains, lawyers, psychologists and educators. In the NZDF, the term specialist is often used to define those who receive their specialist training prior to enlistment. Warfighters often denigrate these people as transitory impostors – traitors who harbour greater loyalty to another profession. It is conceivable this differentiation is less obvious in larger militaries where such members complete their military induction prior to their specialist training. In the NZDF, though, size matters. With so few numbers, it is seldom economical to provide, or even sponsor, specialist qualifications.

It is not just specialist officers who identify with dual professions. Many warfighters and second-tier support branches also have recognizable civilian identities. Two obvious examples include pilots and engineers. A simple test is what job title they give when meeting someone from outside the military. Those who perform uniquely military roles are more likely to describe their occupation by the service they belong to, while those who identify with an alternate profession will give that first and then maybe qualify it with the military context. This simplification might be done for ease of understanding by the civilian, but might also be suggestive of how the person identifies. Dual profession territorial force personnel are probably even more likely to relate to their non-military profession first, once again a reason why warfighters tend to denigrate them as perceived impostors in the Profession of Arms. But there are many other factors creating a delta between the espoused and in-use military culture.

It would be difficult to claim there is no delta between the espoused and in-use cultures. Espoused cultures are usually idealistic and inspirational. While many stewards would claim they are realistic, this is probably more true at the micro and meso levels (individual and small team) but harder at the macro (collective organizational). This is made harder by the fact that pluralist militaries are made up of overlapping sub-cultures, each with their own normative behaviours. Even when combined at the macro level, both types of culture are multi-dimensional and without obvious metrics. Individuals or groups may be strong in one area but weak in others.

Measuring the delta is equally challenging because it is difficult to compare different aspects of a culture.[34] When comparing two similar people, is it possible to say one has more integrity than the other? Does one major breech of trust outweigh years of daily loyalty? And then there is the issue of revealing masks and façades. How do we really know when a member has truly internalised the organization's values? There are of course many other important values (virtues) beyond the espoused bumper sticker ones. These should also be factored in somewhere.

Just as higher ethical principles guide moral dilemmas, the key tenets of an espoused culture are instructive of broader expectations. The natural leaders who influence the workplace in-use culture often realize they are shaping the local culture but not necessarily with the espoused culture, or delta, in mind. Normative changes to sub-cultures are typically natural responses to a variety of sociological factors. These can include both the social influences of wider society as well as the unique military environment. On operations, for example, the culture of a deployment is typically influenced by not only the obvious change in location and task, but also increased effort made by commanders and planners to focus the mission. The cultural delta typically increases the further, or longer, a person or unit is from operations. While regular exposure to operational activity has become a feature for many, there are others in the military that need artificial injections.[35]

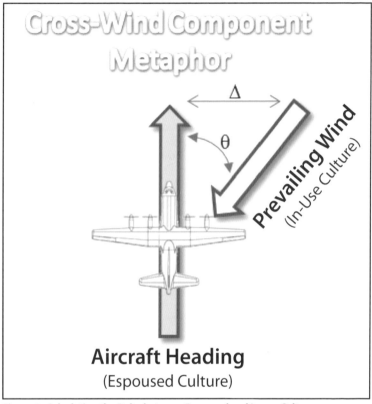

Figure 5: Calculating the Delta between Espoused and In-use Culture

The difference between the in-use and espoused culture can be thought of in the same way aircrew calculate the cross-wind component on landing (Figure 5). Every aircraft has specified limits on the acceptable cross-wind component (Δ) during a landing. This is calculated through simple trigonometry based on the angle of variation (θ) and the strength of the prevailing wind. When the metaphor is used to explain cultural differences, both the strength and variation of the in-use culture (relative to the espoused culture) are important. Furthermore, larger aircraft can sustain greater cross-wind components. For the military, this aspect refers to the history, depth and strength of the espoused culture as an accepted benchmark. The key message of this analogy is that espoused and in-use cultures need not be identical, but the absolute difference (Δ) needs to be within limits. The other important point is that several factors can influence what those acceptable limits are. With military aircraft, for example, the condition of "operational necessity" gives the captain authority to exceed peacetime rules in order to achieve the mission. In the cultural alignment version, this is analogous to local (and even strategic level) commanders accepting an in-use culture that is significantly at variance to the espoused one if it serves a greater purpose.

RAISON D'ÊTRE

Perhaps the most significant influence on the professional ideology, though, is the understanding of a higher purpose. As with the opening parable of the stonemasons, the man who appreciated the ultimate goal was probably far more likely to see the project completed by improvising when necessary and doing whatever it takes to get the job done. He was also more likely to take pride in his work and put the good of society before himself. Military units with low morale could do well to ask how focused their team is on achieving a greater goal.

Operationally deployed units seldom have a low concept of professional ideology. Many commanders lament the absence of military (professional) ideology in peacetime units when they know how quickly the delta disappears on operations.[36] This weak professional ideology in peacetime is often due to a poor understanding of their contribution to the organization's goal. Those who regularly serve on operations will have a clearer understanding of their role in defending the nation's interests.[37] In contrast, personnel more removed from the frontlines will often struggle to appreciate how their work contributes. These second and third echelon workers will typically identify with Moskos' notion of occupationalism. The key message here is for commanders of all levels to reinforce the contribution of even the most menial tasks. But this alone might not be enough.

Perhaps the NZDF's greatest challenge in enhancing the professional ideology is promoting its *raison d'être*. Ideally this would begin by improving the NZDF's public profile as a positive contributor to New Zealand society.[38] Unlike most of our allies, New Zealand's public has little concept of what our military does or why it is even needed. A recent study in fact found 84% of New Zealanders believe the NZDF is incapable of defending itself if attacked.[39] Yet this is only the superficial level. Unfortunately, the study did not explore the public's understanding of the NZDF's contribution to wider national interests such as economic prosperity through trade or even global, regional and personal security both in New Zealand and abroad. Had the study asked such questions, it is likely most respondents would have been oblivious.

Chapter 5

To be fair, the military's contribution to national power is complex and poorly understood, even within the military itself. Without overt aggrandizing, a subtle, yet planned promotion of the Defence Force could have a cascading positive effect not only on recruiting and retention, but also professional ideology within the supporting echelons. Implementing such a communications plan could prove to be the most significant factor in improving professional ideology in the NZDF.

SUMMARY

This chapter has explored professional ideology from a New Zealand perspective. It began by defining ideology as a concept capturing both what and how an organization achieves its goals. The notion of professional ideology was then contrasted with market and bureaucratic ideologies to establish an understanding of service before self and pursuit of a higher goal. This in turn has linkages with the notions of a profession, and in the military's case, the Profession of Arms.

The state of professional ideology in the NZDF is assessed as good, but with room for improvement. Effort is being made to develop and promote specialized knowledge and the culture. Leaders at all levels continue to maintain an acceptable delta between the espoused and in-use culture. Their endeavours, though, have been largely as a result of intuitive responses. The formal education system is only now beginning to appreciate the gap in this subject area.

Enhancing the NZDF's professional ideology faces many challenges. Teaching values on courses has traditionally relied on role model instructors and intuitively conceived activities. Deliberate guidance is increasingly standardizing the development in a cohesive and progressively structured manner. Other challenges still exist with the difficulty in measuring and observing culture. Ultimately, this chapter contends the most important tenet of professional ideology is appreciating the higher purpose. If the NZDF can clearly convey its contribution to New Zealand society to both the public, and members of the Defence Force, the degree of professionalism will increase.

> E tū ki kei o te waka,
> kia pakia koe e ngā ngaru o te wā

Stand at the stern of the canoe and feel the spray of the future biting at your face

ENDNOTES

1. The term professional, in this context, can have three meanings. It might simply refer to being paid (vice amateur), it could mean a sense of great attention to detail and skill, or it could mean belonging to a profession (as opposed to having a trade or craft). Professional ideology, on the other hand, is far more complex than the stonemason parable suggests and is therefore explored in more detail later.

2. Samuel Huntington juxtaposes *applying violence* with *managing violence* when he distinguishes NCOs and soldiers from officers. He uses this distinction to justify his claim that only officers belong to the Profession of Arms. This now-dated attitude is not endorsed by the NZDF.

3. Bill Bentley, *Professional Ideology and the Profession of Arms in Canada* (Canada: The Canadian Institute of Strategic Studies, Brown Book Company, 2005).

4. In the NZDF, these are known as output classes.

5. Daniel Bell, *The Cultural Contradictions of Capitalism*, quoted in Bentley, *Professional Ideology*, 52.

6. Larry Johnstone, *Ideologies: An Analytical and Contextual Approach* (Peterborough, ON: Broadview Press, 1966), cited in Bentley, op. cit., 52-53.

7. Eliot Freidson, *Professionalism: The Third Logic* (Chicago: Chicago University Press, 2001).

8. Some of the leading commentators include Huntington, Janowitz, Moskos and Bentley.

9. Bentley, op. cit., Chapter 2.

10. The zero-sum argument is not universally agreed upon. Some commentators question whether in fact the three dimensions could be complementary.

11. Bentley, *Professional Ideology*, 54.

12. Charles Moskos and Frank Woods, *The Military. More Than Just a Job?* (Washington: Pergamon-Brassey's, 1988).

13. Space precludes a full exposition of the I/O construct, which flourished in the 1980s and 1990s, but numerous publications from that era explored it in great detail. The general concept was that some members of the military placed the institution's needs above their own (institutionalism). Those who see the military as just a job (occupationalism) had no concept of unlimited liability and simply use the organization for personal gain. The apparent rise of occupationalist thinking in militaries resonated with many concerned leaders who believed the Profession of Arms was under threat. The phenomenon is now recognized as a wider societal issue and is often captured with labels such as Generation X and Y. In the context of this topic, occupationalism is akin to market ideology.

14. "The NZDF Joint Vision Statement", unpublished internal NZDF document (Strategic Commitments and Intelligence (SCI) Branch, 2007).

15. J. Codd, "Education policy and the challenges of globalization: Commercialisation or citizenship?" in J. Codd ed., *Education policy: Globalization, citizenship, and democracy* (Southbank, Victoria: Thomson Dunmore Press, 2005), 7.

16. Ibid., 9.

17. A. Fritzson, L. W. Howell Jr. and D. S. Zakheim, "Military of Millennials," Resilience Report, *Strategy and Business*, 10 March 2008, 49.

18. Interestingly, there appears to be conflicting evidence of this. While the thesis is strongly supported by warrant officer of the air force, the warrant officer of the navy reports observing the opposite with new recruits.

19. The NZDF Joint Vision Statement, loc. cit.

20. Ibid.

21. Ibid.

22. Ibid.

23. The RNZN Values were initially derived by a core panel but were then circulated to the wider navy before being finalized.

24. This controversial chapter in the NZDF's history is far more complex than has been shown here and is remembered differently by those involved. The fact that it is so controversial is in itself instructive.

25. For a useful introduction on wicked problems, see the overview on Wikipedia.

26. The *strategic corporal* is a concept where any member of the military can have a strategic effect due largely to the globalisation of technology and media such as CNN and the internet. For more on this, see David Schmidtchen, *The Rise of the Strategic Private : Technology, Control and Change in a Network-Enabled Military* (Duntroon, A.C.T.: Land Warfare Studies Centre, 2006).

27. Brent Davis, Dennis Sumara and Rebecca Luce-Kapler, *Engaging Minds* (3rd Edition) (Routledge, NY: 2008), 76-77.

28. For more on the influences see the various works of both Lawrence Kohlberg (stages of moral reasoning) and David Krathwohl (taxonomy of educational objectives in the affective domain).

29. Emerging and revised military concepts (for example, global terrorism, 3 Block War, Counter-Insurgency Operations and 4th Generation Warfare) highlight the continual evolution in specialist knowledge within the Profession of Arms.

30. Robert Walker, *The Professional Development Framework: Generating Effectiveness in Canadian Forces Leadership* (Kingston, ON: Canadian Forces Leadership Institute Technical Report 2006-01, 2006).

31. For more on this discussion from a New Zealand perspective, see: Murray Simons, *Professional Military Learning: The Next Generation of PME in the New Zealand Defence Force* (Air Power Studies Centre, Canberra, 2005).

32. Lyle Spencer and Signe Spencer, *Competence at Work: Models for Superior Performance* (Toronto, ON: John Wiley and Sons, 1993), 11.

33. Empirical evidence shows most recruits display higher levels of institutionalism than those who have been serving for several years. For more on this from a New Zealand perspective, see: Murray Simons, *Identifying the True Military Factor in RNZAF Training* (Unpublished M.Ed thesis, Canterbury University (NZ), 1997).

34. Psychometricians driven by market ideologies would no doubt contest this assertion.

35. Examples of these include renaming all branches to include the name operational in them, the introduction of an all-compulsory week-long "core military skills" course for all uniformed members, and the creation of "Air Power Friday," when those on support bases wear their operational clothing once a week, regardless of their current employment.

36. Several senior personnel commented on this when interviewed for the chapter. The current chief of air force, for example, proactively encourages all levels of command to "train as we fight." By this he means employ *mission command* by trusting subordinates and exploiting innovation at all times – not just on operations. The tenets of professional ideology should be valued and employed at all times.

37. It is worth noting that Operational Tours tend more toward a professional ideology because of the *relative* reduction in bureaucratic ideology (including secondary appointments and non-core distractions) imposed at home. There also tends to be less market ideology influences on members of the deployment while they are removed from the distractions of society and are highly focused on their mission.

38. General Hillier (ret'd), during his tenure as CDS of the Canadian Forces, was renowned for his efforts in raising the military's profile in society. The senior leaders of the NZDF would do well to study his approach.

39. Dale Elvy, "Defence: exploring the silent consensus," *New Zealand International Review*, May/June 2008, 33(3), 23–26.

CHAPTER 6

PROFESSIONAL IDENTITY
IN THE SWISS ARMED FORCES:
A CONTRADICTION IN TERMS?

Dr. Stefan Seiler *

Questions concerning professional identity in the Swiss Armed Forces are both relevant and politically delicate. To fully appreciate discussion surrounding professional identity in the Swiss Armed Forces, one has to take into account the Swiss understanding of the military and of the federal state (i.e., the Swiss military has to be seen in its social and political context). Thus, the first section of this chapter provides a brief overview of the organization of the Swiss military and an understanding of the federal state. The second section considers the relevance of the professionalization debate for the Swiss Armed Forces. The next section identifies defining elements of the professional identity of Swiss career officers. The fourth and final section synthesizes the previous points into a comprehensive model of professional identity for career officers in the Swiss Armed Forces, which is intended as a starting point for further discussion. The model builds on the principle of the militia as well as on the specific tasks of career officers. This chapter concludes with a summary and discussion of the future relevance of professionalization in the Swiss Armed Forces.

THE MILITIA ORGANIZATION OF THE SWISS ARMED FORCES

Military service in the Swiss Armed Forces is compulsory for all Swiss male citizens.[1] Although not compulsory, female citizens can volunteer.[2] Following six months of basic training, recruits are assigned to an operational unit and serve three weeks each year over the course of six years. On average, compulsory service totals about 300 days. Approximately ten percent of conscripts complete their 300 days consecutively. Although the subject of debate, the Swiss parliament has recently decided not to increase the percentage of this type of service on the grounds that it would violate the principle of having no

---

* The ideas expressed here represent the author's point of view and not the official view of the Swiss Armed Forces.

standing military troops, which is founded in the Swiss Constitution. Operational units complete their three weeks of annual service at different times of the year. This ensures that one or several military units are on duty at any given point, thus ensuring the rapid availability of personnel in case of unexpected crises (e.g., natural disasters). At present, the Swiss Armed Forces consists of about 120,000 active duty members and 80,000 members in the reserve.

This system is called the "militia system." It stands in direct opposition to a system of professional soldiers because militia soldiers are Swiss citizens who hold civilian jobs. They join the armed forces and complete their military service for three weeks each year. The militia principle applies to all ranks. Upon being accepted, conscripts who choose a career as an officer or non-commissioned officer undergo a short basic training course followed by occupational training. This is followed by a period of employment in a basic training formation as an instructor and eventually they lead these formations. Annual duty in an operational unit in a leadership position often follows. A first lieutenant, for example, may serve as a platoon leader in a company, requiring three weeks of service each year with troops. Given a successful track record, officers who are interested in continuing their military career can apply for a company commander training course to become company commanders or complete a staff training course for eventual employment as a battalion staff officer. The same procedure is adopted for future battalion and brigade commanders and staff officers at the brigade level. Operational units up to brigade level may therefore fall under the command of militia officers who, for the better part of the year, hold civilian jobs but spend a couple of weeks each year as military leaders, instructors and educators of their subordinates during operational unit training. This system is also applied in operational units that support civilian authorities in fulfilling their tasks (e.g., at the annual World Economic Forum in Davos, where a large number of conscripts support the police services). Their militia officers lead these troops.

Meanwhile, the number of career officers in the Swiss Armed Forces is relatively small. Of the 20,000 officers, only three percent (about 630) are career officers and of the 35,500 non-commissioned officers, only four percent (about 1,400) are full time; the rest are militia cadre.[3] The numbers demonstrate that the degree of professionalization within the Swiss Armed Forces is rather low.

THE RELEVANCE OF THE PROFESSIONALIZATION DEBATE

The current organization of the Swiss Armed Forces seems to resist discussion about professionalization. It is argued that a militia, by definition, is managed by citizens, not by professional soldiers. This philosophy makes a high degree of professionalization undesirable, if not impossible, because to fully professionalize the armed forces would jeopardize the principle that it be an army of the citizens. The militia rationale pervades other domains of Swiss public life as well. The national parliament, for example, is made up of "militia politicians" (i.e., members of the Swiss National Parliament are not full-time politicians). They have a civilian profession and are only reimbursed for their political activities for the few weeks they meet during parliamentary sessions each year and for the additional time they spend in special parliamentary working groups. Hence, they are not paid a regular salary or committed to a fixed term of service. The militia organization mirrors the Swiss' understanding of the federal state, in which the citizen wields authority over federal institutions and where citizens refuse to relinquish too much or absolute power to the government. Hence, the majority of Swiss officers are not directly involved, nor necessarily interested, in the professionalization debate. A militia officer is a civilian citizen pursuing a civilian job who, in addition, is a military officer. Militia officers, therefore, do not qualify as "professionals" in their capacity as officers, especially if the discussion of military professionalism is based on a strict definition as enunciated by Huntington.[4] According to Huntington, professionalism is founded on three criteria: expertise, responsibility, and corporateness. Expertise refers to the specialized knowledge of the professional practitioner, gained through ongoing extensive study of the profession and practical experience in the field. Militia officers do not experience either: limited study of their duties as officers is offered and this is coupled with limited opportunity to obtain practical leadership experiences (i.e., actual missions).

The second criterion of professionalism is responsibility. For Huntington, responsibility refers to the provision of services that are beneficial to society. It requires a profession to regulate its members by enforcing ethical codes. The responsible individual derives intrinsic motivation from a passion for his or her profession and is committed to the state by a "…sense of social obligation to utilize this craft for the benefit of society."[5] This sense of social obligation and commitment plays an important role for both career and militia officers, as do codes and ethics. Further, militia officers are required to follow the service regulation manual (DR04) and are subject to military law. Thus, this criterion is fulfilled to some degree by both career and militia officers. Of note, however, there are no special ethical codices for members of the Swiss Armed Forces. Although the DR04 mentions certain rules, duties and behaviour, it does not specify a code of conduct.

Chapter 6

Huntington defines the third criterion of professionalism, corporateness, as the professionals' "…sense of organic unity and consciousness of themselves as a group apart from laymen."[6] This criterion, the sense of belonging and the necessity to set yourself apart from other groups, concerns militia and career officers alike. It is applicable to officers in respect to soldiers, but also to militia officers vis-à-vis career officers: militia officers go through a clearly structured training sequence, at the end of which they are promoted; career officers undergo additional training and receive the career officer diploma.

Unlike their militia counterparts, career officers have an ever-increasing stake in the professionalization debate. Though there are few of them, the hazy nature of their professional status within the current system makes it difficult to recruit future career officers. The position of their profession within the armed forces and society remains unclear, even though they are widely considered the "backbone" of the military because of their significant contributions to maintaining standards and fostering progress. The role of career officers is bound to assume even more importance in the future: faced with rapid changes to national security and defence strategy agenda, the current system will quickly reach its limits and increasingly become reliant on professional and semi-professional support. The carriers of this professional support must have a well-defined professional identity; if not, the system is likely to lose its stability. The current system struggles particularly with the growing demand for soldiers for international UN peacekeeping operations abroad. For example, since sending militia soldiers abroad is only permitted on a voluntary basis and for limited contract periods unrelated to compulsory service, it makes recruiting sufficient personnel for such operations a significant challenge. On top of this, it is nearly impossible to accommodate the training requirements for interoperability within the brief time span available to operational units. Units also face a similar problem (i.e., limited time) when training on new weapon systems. As a result, Switzerland is currently in a debate over the need to increase professionalization within the armed forces while preserving the identity of the militia organization.[7]

Since the end of the Cold War and the virtual disappearance of a military threat to the country's borders, the relevance of the military has been questioned. Over the past few years, growing skepticism about the purpose of the military and a general deterioration of working conditions within the military have worsened the situation of career officers. This has also hampered recruitment. A clear professional identity would no doubt strengthen the profession by improving the perception of the professional military from within and external to the armed forces. A murky identity, on the other hand, will make the Swiss Armed Forces an easy target for criticism and possibly a plaything for politicians.

Further, in the eyes of many Swiss, the status of the militia officer has ceased to be a platform for career opportunities in a civilian profession. Many executive posts in the economy and in politics were once held by the same people who held key positions in the military. Due to increasing globalization, however, the gaps between the military and civilian sector have widened and the status of a militia officer has lost considerable appeal.[8] Militia officers once benefited from the skills they acquired as well as the social network they nurtured within the officer corps, but the importance of the latter has diminished significantly. The coordination and juggling of civilian employment and military service has become increasingly problematic. The armed forces readily admit that they are having difficulty recruiting new and qualified militia cadre. In response, they have launched several measures to reverse this trend (e.g., marketing activities, offering better financial compensation for militia training programs, and the accreditation and/or granting of civil diplomas for military training). The question is whether a higher degree of professionalization could improve the reputation of the military as a whole.

DEVELOPING A PROFESSIONAL IDENTITY

Professionalization in the Swiss Armed Forces must be discussed in conjunction with the militia philosophy. Even though career officers are professionals, their professional identity should not deviate greatly from the fundamental notion of the militia system (i.e., the *citizen in uniform*). Professional identity should not be exclusively limited by the militia principle, the general mission of the armed forces, or the traditional tasks of the officer, because in doing so, their identity is then indistinguishable from that of the militia officer. Instead, the militia organization and the mission of the armed forces should serve as the foundation on which the identity of the career officer rests. One of the supporting pillars building on this foundation is the general and shared understanding of leadership. The principles introduced so far are still valid for both militia and career officers. The next two pillars distinguish the career officer from the militia officer. They include, first, the tasks carried out specifically by career officers and, second, a mission statement tailored to military career personnel. Before these two pillars are further explained, the foundation and the first pillar will be illustrated.

Chapter 6

The Foundation of Professional Identity in the Swiss Armed Forces

The foundation of professional identity in the Swiss Armed Forces consists of two components, the first as previously discussed, being the organization as defined by the Swiss Constitution. The second is the mission of the armed forces, as described in the DR04, as well as in the standards of leadership, training, discipline, day-to-day service management, and the rights and duties of military members.[9] As chapter 2 of the DR04 indicates:

> The Swiss Confederation seeks peace and the preservation of its liberal democratic foundation. It wants to defend the state territory in case of military conflicts and protect the people and their existential basis. It wants to contribute actively to securing world peace and stand up for human dignity.[10]

The DR04 further mentions the need to co-operate within the international community of states and defines the military as one instrument toward this end. Other instruments include foreign policy, civil protection service, economic policy, state logistics, state protection, the police, and communication and information services.[11] Included in military tasks – tasks in accord with the Swiss neutrality policy – are:

a)   helping to prevent war and to secure peace;

b)   defending Switzerland and protecting the population;

c)   contributing to peace on an international level; and,

d)   supporting domestic or foreign civil authorities whose means are insufficient to cope with severe threats and catastrophes.[12]

These tasks legitimize the military as an institution and define its mission range. They, along with the militia organization, represent the foundation of the career officer's professional identity. It is crucial that officers be able to identify with these tasks since they provide both the democratic rationale and the mission profile of the officer's actions: prevent war, secure peace, defend Switzerland and the Swiss people in case of war, contribute to international peace, and support civil authorities at home and abroad to cope with catastrophes and to maintain inner security. The list is final and missions deviating from this task list are illegitimate. If a legitimate mission is assigned to the armed forces, officers must apply their skill and effort to the service of the military (and hence the state) to accomplish their mission with the means available.

Leadership Ideology in the Swiss Armed Forces

The DR04 offers certain guidelines that concern the kind of behaviour expected of soldiers in the pursuit to achieve mission goals. Order and obedience are considered the clearest expression of military leadership. The DR04 also contends, however, that leadership in operations consists of much more than order and obedience. Comprehensive military leadership includes setting goals, making decisions, handing out tasks, co-operating with others, avoiding or settling conflicts, motivating subordinates and showing a concern for their well-being. Subordinates adhere to the principle of obedience, but depending on mission requirements, are expected to act independently, with discipline and on their own responsibility. They should actively forward information and work effectively with others. Meanwhile, superiors must define clear aims and have the courage, trust in and respect for their subordinates and the willingness to delegate or provide subordinates with freedom of action to best facilitate mission success. Since all military superiors are subordinates in turn, military conduct demands discipline, autonomy, self-responsibility and teamwork on all levels of the hierarchy. In this sense, military leadership goes well beyond the principle of order and obedience. It is about:

- leading with defined aims;

- proactive thinking;

- commitment;

- taking responsibility on all levels;

- discipline;

- information distribution and communication adequate to the addressee;

- exemplary conduct; and,

- co-operation and motivation.[13]

These leadership maxims apply to both militia and career officers and represent the first pillar of a Swiss professional identity that is hinged on the foundation of the militia organization and missions. It is obvious that the leadership ideology of the Swiss Armed Forces differs from the soldier ethos described by Huntington, which is determined primarily by obedience and loyalty.[14] For Huntington, the military is an instrument of the state and soldiers are required to fulfill their tasks as efficiently as possible. This train of thought seems logical enough, but may carry the risk of soldiers eschewing moral responsibility for their actions. Numerous examples demonstrate that the call for obedience and loyalty can be justified only within certain limits because, eventually, actions cannot be separated from the person who commits them.[15] A soldier must base his or her actions on the mission, but often "there are unique situations, contexts and problems, and these do generate unique moral demands."[16] Obedience and loyalty do not free soldiers from moral responsibility. Thus, the leadership ideology of the Swiss Armed Forces rests on a more differentiated view of the interaction between superiors, subordinates, organization and mission, and aspires to a modern, adult-oriented concept of leadership that integrates both the principle of obedience/loyalty and the principle of self-responsibility and independence.

Excursus 1: The Swiss Armed Forces do not have a code of conduct or a virtue and value catalogue as found in many other armed forces, mainly because the idea of a "citizen in uniform" needs no justification other than the democratically legitimized mission of the military along with the leadership ideology mentioned above. Citizens in uniform are also civilians, and as such, have no desire or need for a distinctive military subculture.

Yet codes of conduct and value catalogues can also be questioned with respect to their content. Seiler lists several criticisms of codes of conduct and catalogues of values, e.g., their limited scope, the unsystematic prioritization of certain values over others, the lack of clarity as far as the effects of such catalogues are concerned, or the problem of disregarding the situational context of decision-making. In addition to the close interaction between civilian and military culture, these general criticisms of codes of conduct are the reason why the professionalization debate within the Swiss Armed Forces does not seem to merit a plea for such codes. Our discussion should focus on "lived leadership" that balances responsibility, fairness, care, integrity and situational context, rather than a set of codes and values.

Specific Tasks of the Career Officer

Career officers perform certain tasks that militia officers do not. These tasks form the second pillar of the career officer's professional identity. Successful performance for a career officer necessitates a high degree of professional knowledge, skill and experience:

* instructor, leader and educator during basic training and for militia cadre;

* commander of operational units;

* staff officer; and,

* military expert in the areas of training, mission planning and mission command, doctrine, armament supervision and international relations.[17]

As evidenced by this list, the spectrum of tasks performed by career officers is quite wide, ranging from teaching militia troops to leadership tasks in operational units to consulting/project management related to military, security and foreign policy. In a complex global environment, these tasks call for professional executives and experts. Career officers have to be professional instructors and credible leaders, both domestically and internationally; they command operational units at home and abroad and are deployed in international staff functions. In their capacity as expert consultants, they must acquire a thorough knowledge of, amongst other areas, the military's mission range, domestic and foreign weapon systems, and security and foreign policy. Therefore, the Swiss Armed Forces cannot avoid the professionalization debate. Even if the number of pure professionals remains small, there are increasingly more and more tasks that can only be performed by a small group of people because the militia cadre lacks the necessary competence and availability. Consequently, the Swiss Armed Forces consider the selection, training and further education of career officers of paramount importance. In addition to successfully completing officer training and a fitness check, all career officer candidates must attend an assessment centre that systematically evaluates their social and intellectual capability. Passing the respective exercises and tests at the assessment centre qualifies candidates for further training. Depending on their academic background, accepted candidates go through one to three years of selected academic coursework at the Swiss Military Academy at the Swiss Federal Institute of Technology (ETH) in Zurich. Upon completion, candidates receive the career officer diploma and in certain programs, a bachelor's degree in political science. In addition to courses taken with militia officers, career officers complete special courses, spanning several weeks, as part of their preparation for more complex functions. These courses are to some extent coordinated with the ETH. Participants of the "Commanders Training III" course, for example, which is designed for future brigade generals, are granted a full academic title at the end of their studies (a Master's of Advanced Studies in Security Policy and Crisis Management). This and many other examples illustrate the significant role that further education plays in the professional life of a career officer.

The professionalization of the militia system is not a contradiction in terms as long as the professional component forms an integral part of the militia philosophy. Professionals bring the necessary competence for complex, specialized, long-term tasks (e.g., the training of militia cadre, commanding troops in missions abroad, consulting in military and security policy domains, and leading large training formations, etc.). In turn, the militia focuses on missions appropriate to its level of expertise (e.g., leading militia formations in training exercises involving several units, supporting civil authorities in subsidiary missions, and preparing for defence in times of war, etc.). Whether the integration of the professional component will work in practice remains to be seen. There is always a danger of creating a two-class army in which career officers carry out all the important tasks and the career of the militia officer is no longer an attractive choice.

A Mission Statement for Career Personnel

In order to do full justice to career personnel, the Swiss Armed Forces have developed a mission statement describing the particular values and norms of this professional category.[18] According to the mission statement, the central task of career personnel is the training, operational readiness and the maintenance and evolution of the militia. This is yet another expression of how militia and military professionals are inseparable. The statement also asks career personnel to keep an open mind with respect to social and technological developments while ensuring the passing on of the tradition of the corps spirit to new soldiers and officers. Apart from such general tasks, the statement also mentions certain specific requirements. For example, a career soldier serves as a role model by virtue of exemplary and credible conduct; he or she is characterized by competence, versatility, flexibility and interoperability. A sub-section of the statement is devoted to the topic of mission and career choices in that the professional development of career personnel should be systematic, selective, competitive, transparent and co-operative.

Seven guiding principles have been derived from the mission statement. The principles are meant to guide and motivate career personnel as well as help them develop their professional identity within their corps. The principles are as follows:

◆ We lead prudently and are goal-oriented, and we show concern for our subordinates!

◆ We lead by example!

◆ We are flexible with regards to content, time and location!

◆ We systematically pursue further education!

◆ We train our soldiers competently!

◆ We are versatile in our function as military experts! and,

◆ We lead missions in cooperation with civil and military partners![19]

These guiding principles for career officers complement the general principles of military leadership and provide the basis for a catalogue of competencies for career officers: a career officer is a prudent and goal-oriented leader, who leads by example, who can be deployed flexibly with regard to time and location, who can be placed in his or her capacity as instructor, leader, expert, and who works well with civilian and military partners. This profile follows from the specific tasks of the career officer previously mentioned. A career officer displaying the respective conduct and competence is able to accomplish his or her tasks, which is why the mission statement is the third pillar of the model of professional identity (see Figure 1 on page 108).

Excursus 2: The importance attributed by the mission statement to career development shows how the military views itself as a modern employer who wants to be a fair, transparent and competitive partner for their employees – a crucial quality asset in an age where competition on the executive job market is harsh. In a global and individualized society, career officers perceive their profession primarily as a job rather than a calling (illustrated by the increasing number of resignations by career officers over the past few years). This tendency must be reversed by repositioning the career cadre within society and the military on the one hand, and by creating good working conditions on the other. Lack of career officers aside, it is reassuring to know that officers who resigned were able to find new and attractive positions in the economy or in the administration sector, so their prospects on the job market in other professional arenas appear to be intact.

Chapter 6

A MODEL OF PROFESSIONAL IDENTITY OF CAREER OFFICERS IN THE SWISS ARMED FORCES

The final section of this chapter develops the model of professional identity for career officers in the Swiss Armed Forces. The model incorporates the militia principles particular to the Swiss Armed Forces, while illustrating the specific aspects of their professional identity. The foundation and the three pillars of the model are shown in Figure 1:

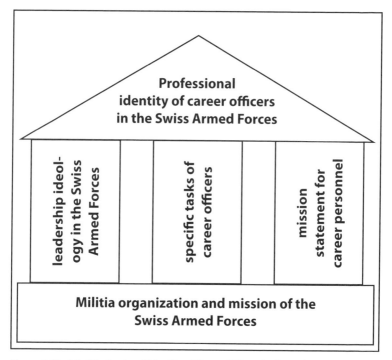

Figure 1: Model of Professional Identity of Career Officers in the Swiss Armed Forces.

The Foundation: Militia Organization and Mission

The foundation of a career officer's professional identity is the constitutionally regulated organization of the armed forces as a militia and their mission. All actions are based on these two elements since they determine the form of the armed forces and the mission range. They provide career officers with a rationale for the existence and purpose of the military on which to build their professional identity.

The First Pillar: Leadership Ideology of the Swiss Armed Forces

The first pillar that rests on the militia organization and mission foundation is the leadership ideology of the Swiss Armed Forces. According to this ideology, military instruction should use modern means and methods of training and instruction and that leadership and education should be adult-based. Discipline and loyalty are considered central aspects of instruction, but independence, self-responsibility and commitment on all levels are equally important. The leadership ideology forms the basis for the responsible conduct of each individual soldier aiming to accomplish his or her mission, and if necessary, at the cost of their own life – an aspect that distinguishes the armed forces from almost all other organizations (excepting the police and fire department, and some private security firms). Military missions require people to stay focused, act responsibly and be goal-oriented in extreme and life-threatening situations. Such situations demand a good deal of competence, discipline, self-responsibility and trust in leadership.

The Second Pillar: Specific Tasks of the Career Officer

The tasks that are performed specifically by career officers were previously listed in this chapter. They range from instructor to leadership tasks to consulting and project management in the military to tasks in the security and foreign policy domains. These tasks are the essence of a career officer's professional identity because they distinguish him or her from the militia officer. Why then not use this pillar as the foundation of the professional identity? Several reasons speak against this: first, the Swiss Armed Forces are constitutionally defined as a militia, and the tasks of all members of the armed forces, conscript and professional, have to be anchored in the militia system. Moreover, the function of career officers is to enable militia personnel to accomplish their missions. This is demonstrated in the mission statement that justifies the presence of career personnel by their contribution to the training, operational readiness, maintenance and evolution of the militia. Second, service regulations in the DR04 apply to both career officers (during their entire work period) and militia officers (during service). Third, all career officers fulfill a militia function that most often deviates from their professional function. Hence, most are confronted with the tasks of both the career and the militia officer. As the second pillar contains the most distinctive feature of the career officer's professional identity, it stands in the centre of the model.

Chapter 6

The Third Pillar: The Mission Statement for Career Personnel

The mission statement for career personnel specifies competencies and behaviour that complement the leadership standards covered in the DR04. The additions made by the mission statement focus on the tasks of the career cadre and thus provide the content for the second pillar in the form of task descriptions and practical guidelines. In other words, the mission statement elucidates how the defined tasks are to be accomplished. It determines the actual performance of career officers as it lends their actions a distinct shape vis-à-vis those of non-professional officers. The statement plays a crucial role in creating and maintaining the professional identity of career officers because it offers a concrete picture of how a career officer's actions should manifest in practice.

SUMMARY AND OUTLOOK

This chapter discussed the professional identity of militia and career officers in the Swiss Armed Forces. As illustrated, the professionalization debate in the Swiss military must take into account the militia organization laid down in the Swiss Constitution and address the dilemma of having a professional component within a militia organization. In order to do justice to the militia principle, the model of professional identity presented here takes the organization of the militia and the general mission of the military as its foundation and only then introduces the three pillars of professional identity in the Swiss Armed Forces: leadership ideology, the specific tasks of career officers, and the mission statement for military career personnel. The model establishes a basis for a differentiated discussion of professional identity in the Swiss military. The professionalism debate, although not new, is crucial to the profile of career officers. Their profession is in need of a substantial professional identity that can inspire a sense of self, be used for defining and enforcing professional standards, creating a consistent outward appearance, and positioning the profession of career officers on the job market. These are basic requirements to attract young qualified men and women and to ensure that career officers can be proud members of their profession, both in the military and in society.

ENDNOTES

1. *Schweizerische Eidgenossenschaft: Bundesverfassung* [Swiss Confederacy: Constitution] (Berne: Federal Chancellery, 1999).

2. *Schweizer Armee: Dienstregelement DR04 Reglement 51.2.* [Swiss Army Service Regulation Manual] (Berne: EMZ, 2004), 3.

3. All numbers are approximations. The overview presented here, although listing the essential aspects of how the Swiss Armed Forces function, is naturally an oversimplification and does not offer a comprehensive description of all the details related to the matter at hand.

4. Samuel Huntington, *Soldier and the State* (New York: Vintage Books, 1957).

5. Ibid., 31.

6. Ibid., 26.

7. Karl Haltiner, "Allgemeine Dienst- statt Wehrpflicht: aktuelle Idee oder Schnee von gestern?" [Public Duty Instead of General Conscription: Role Model or Water Under the Bridge?], in Karl Haltiner, Urs Wenger and Silvia Würmli, eds., *Allgemeine Dienstpflicht. Leitbild oder Schnee von gestern?* [Public Duty. Role Model or Water under the Bridge?] (Zürich: Militärakademie an der ETH Zürich, 2007), 4-11.

8. An interesting description of the former functioning of the Swiss militia system can be found in the book *La Place de la Concorde Suisse* by John McPhee (New York: Farrar, Straus and Giroux, 1984). Today, McPhee's description is no longer entirely accepted. The social and political landscape has changed and many who hold executive positions in the economy decide against a career as a militia officer. Soldiers are discharged earlier. These factors contribute to the weakening of the ties between civilian and military life.

9. *Schweizer Armee: Dienstregelement DR04 Reglement 51.2.* [Swiss Army Service Regulation Manual] (Berne: EMZ, 2004), 1.

10. Ibid., 3.

11. Ibid.

12. Ibid.

13. Ibid., 5-7.

14. Samuel Huntington, "The Military Mind: Conservative Realism of the Professional Military Ethic," in M. M. Waking, ed., *War, Morality, and the Military Profession* (Boulder: Westview Press, 1979), 25-50.

15. Stefan Seiler, *Führungsverantwortung. Eine empirische Untersuchung zum Berufsethos von Führungskräften am Beispiel von Schweizer Berufsoffizieren* [Leadership Responsibility. An Empirical Study] (Bern: Peter Lang, 2004).

16. J. Carl Ficarrotta, "Are Military Professionals Bound by a Higher Moral Standard?" *Armed Forces and Society*, 1997, 24(1), 62.

17. *Schweizer Armee: Berufsmilitär - Zeitmilitär. Ein Beruf für unsere Sicherheit* [Military Professional – Temporary Military Professional. A Profession for our Security] (Berne: Stab CdA, PersV, 2006).

18. *Schweizer Armee: Leitbild. Das militärische Personal der Armee XXI* [Mission Statement: Career Personnel of the Armed Forces XXI] (Berne: EDMZ, 2001).

19. *Schweizer Armee: Leitbild. Leitgedanken für das Berufsmilitärkorps* [Mission Statement: Guiding Principles for the Professional Military Corps] (Berne: EDMZ, 2001).

CONTRIBUTORS

Doctor **Bill Bentley** is a retired lieutenant-colonel with 35 years experience in the Canadian infantry. He served in both UN and NATO appointments and was the Canadian Exchange Instructor at the US Army Staff College and the School for Advanced Military Studies. He currently works at the CF Leadership Institute in Kingston, Ontario. He is the author of *Professional Ideology and the Profession of Arms in Canada*. In 2006, he received the Meritorious Service Medal for his contributions to the reform of the CF Professional Development System over the past 10 years.

Lieutenant-Colonel drs. **Coen van den Berg** started his military career at the Royal Netherlands Military Academy in 1983. After serving with the Royal Netherlands Military Engineers, he has worked, since 1994, as a military psychologist (after getting a degree in psychology at the University of Utrecht). He has held postings at the Netherlands Institute for Leadership and the Behavioural Sciences Department of the Royal Netherlands Army, the Veterans' Institute, and is presently working as a lecturer of leadership and military ethics at the Netherlands Defence Academy. His research focuses on the influence of threat on soldiers' mission-related attitudes, psychological support, and cross-cultural competencies during deployment.

Doctor **Elly Broos** has been employed by the Royal Netherlands Navy since 2001 and is currently an assistant professor at the Netherlands Defence Academy. She teaches human resource management, military leadership and ethics. She obtained her PhD in human performance and technology.

Major **Adrian Y. L. Chan** presently heads the Leadership Doctrine and Research Branch at the Centre of Leadership Development, SAFTI. His previous appointment was Head – Training and Doctrine Branch in the Applied Behavioural Sciences Department. He completed his BSc in psychology (2nd Upper) in 1994 at Royal Holloway College, University of London, and is currently a management PhD candidate (with emphasis on leadership development) with the Gallup Leadership Institute, University of Nebraska-Lincoln.

Lieutenant-Colonel, Doctor **Kim-Yin Chan** is currently heading the Continuing Education Office at SAFTI Military Institute. Prior to this, he headed the Leadership Doctrine and Research Branch in the SAF Centre of Leadership Development, SAFTI MI. He is an infantry officer by training and has commanded at the platoon and company levels. He received his BSc in psychology with First Class Honours from the University of London, UK, in 1988, and his MA and PhD in industrial-organizational psychology from

Contributors

the University of Illinois at Urbana-Champaign, USA, in 1997 and 1999. He has published academic research papers in the *Journal of Applied Psychology, Journal of Vocational Behaviour, Personnel Psychology, Multivariate Behavioural Research, Journal of Education and Measurement Research, Pointer: Journal of the SAF,* and *Catalyst: Journal of the Military Behavioural Sciences* in MINDEF and the SAF.

Major drs. **Robbert Dankers** studied public administration and then joined the Royal Netherlands Air Force in 1997. After holding different positions in the field of finance and control, he is now a lecturer for different levels of staff officer courses at the Netherlands Defence College. His topics are business and public administration, with a special interest in change management and leadership.

Lieutenant-Colonel **Gunawan** is a graduate from the Faculty of psychology, the University of Padjadjaran, Bandung, Indonesia (1989). He has a MA in the Psychology of work from the Université Pierre Mendès, Grenoble, France (1995). He also graduated from the Indonesian Armed Force Officer's School (1986), the Indonesian Army Staff and Command School (2003) and the Indonesian Armed Force Strategic Intelligence Course (2006). He currently heads the psychological laboratory at the Psychological Service of the Indonesian Army. His main interests are in psychological operations, assessment and selection.

Lieutenant-Colonel **Eri Radityawara Hidayat** has a Bachelor's of Science degree from the University of Wisconsin, Madison, USA (1986), a Master's of Business Administration from the University of Pittsburgh, USA (1987) and a Master's of Human Resource Management and Coaching from the University of Sydney, Australia (2005). He is a graduate of the Indonesian Armed Force Officer's School (1990), the Australian Military Familiarization Course (2004), the Indonesian Army Command and Staff School (2006) and the Netherlands Defence Orientation Course (2008). He is currently the Head of Planning and Budgeting at the Psychological Service of the Indonesian Army. His main interests are in leadership development and military history.

Kwee-Hoon Lim is currently a field psychologist at the SAF Centre of Leadership Development at the SAFTI Military Institute. She was previously a commissioned SAF officer and has held appointments in an infantry brigade, the SAF Signal Formation, MINDEF Education Department and the Applied Behavioural Sciences Department in MINDEF. Her current work involves research in values and values-based leadership in the SAF. She received her BA in psychology, graduating with honours from the University of Calgary, Canada, in 1989.

Lieutenant-Colonel **Dick Muijskens** is head of a training unit of the Royal Marechaussee at Apeldoorn. In 1985, he finished his basic military training and worked in a variety of operational functions. In 1996, he finished his training at the Netherlands Defence Academy and worked as an officer at the headquarters of the Royal Marechaussee in The Hague and with the Central Investigation Unit in Utrecht. In 1998, he completed studies as an environmental engineer and in 2005 he obtained a master's degree in politics of administration and a post-master's degree in management of organizations.

Lieutenant-Commander drs. **Ineke R.J. Dekker-Prinsen** joined the Royal Netherlands Navy in 1981 and became an officer. She studied work and organizational psychology and is currently working at the Netherlands Defence Academy. Her focus is on leadership and military ethics.

Doctor **Stefan Seiler** is currently the Head of Leadership and Communication Studies at the Swiss Military Academy at ETH Zurich. He studied at the University of Fribourg and the University of Leeds and graduated with a PhD in educational psychology from the University of Fribourg. His research interests include leadership, intercultural leadership, leadership ethics, moral decision-making and conflict management. Prior to his appointment at the Swiss Military Academy, he worked at Credit Suisse in Zurich and in New York as a member of senior management in the human resources department. His responsibilities included global restructuring and implementation projects in America and Asia. Dr. Seiler serves in the army in the rank of major (militia officer) and is a member of the military science workgroup in support of the chief land forces. He was previously a company commander, followed by chief of staff and deputy commander of a pioneer battalion.

Squadron Leader **Murray Simons** is currently posted to the Headquarters of the New Zealand Defence College as the Staff Officer Professional Military Development. His academic qualifications include a Bachelor's of Science in psychology from Auckland University, a Diploma of Teaching (Secondary), a Master's of Education from Canterbury University, a Master's of Management in defence studies from Canberra University, and a Master's of Arts in strategy and policy through the Australian Defence Force Academy (University of New South Wales). He is currently writing his doctoral dissertation on 'Hidden Learning in Professional Military Education'.

Colonel **Sukhmohinder Singh** is a commando officer by vocation. He currently heads the SAF Centre of Leadership Development at SAFTI Military Institute. He has commanded an infantry battalion and brigade, served as the Head of SAF Advanced Schools and the Commander of Army Advanced

Contributors

School. He was instrumental in developing the SAF's strategy for enhancing training and development among its officers in its transformation towards a "3G SAF" concept. He has previously published in the *Pointer: Journal of the Singapore Armed Forces*. He holds a BA in history and political science from the National University of Singapore.

GLOSSARY

3G SAF	Third Generation SAF
9/11	11 September 2001
AAR	After Action Review
C3I	Courage, Comradeship, Commitment, and Integrity
CBHRM	Competencies-Based Human Resources Management System
CDA	Canadian Defence Academy
CDF	Chief of Defence Force
CF	Canadian Forces
CFLI	Canadian Forces Leadership Institute
CLM	Command, Leadership, Management
COC	Code of Conduct
CV	Core Values
DEPJUANG	Departemen Kejuangan (Fighter Department)
DISBINTALAD	Dinas Pembinaan Mental Angkatan Darat (Mental Guidance Service of the Army)
DISPSIAD	Dinas Psikologi Angkatan Darat (Psychological Service of the Indonesian Army)
DM	Decision-Making
DR04	Swiss Service Regulation Manual
ETH	Swiss Federal Institute of Technology
EVC	Eerder Verworven Competenties
FPS	Flexible Personnel System
GOLKAR	Golongan Karya (Functional Group)

Glossary

IMLA	International Military Leadership Association
I/O	Institutionalism-Occupationalism
KASAD	Kepala Staf Angkatan Darat (Army Chief of Staff)
KNIL	Koninklijk Netherlands-Indische Leger (Dutch Colonial Army)
KODIKLAT	Komando Pendidikan dan Latihan Angkatan Darat (Education and Training Command of the Army)
KOTER	Komando Territorial (Territorial Command)
MABESAD	Markas Besar Angkatan Darat (General Headquarters of the Army)
MP	Military Police
NCM	Non-Commissioned Member
NKRI	Negara Kesatuan Republik Indonesia (Unitary State of the Republic of Indonesia)
NSF	Full-Time National Servicemen
NZ	New Zealand
NZDF	New Zealand Defence Force
ORNS	Operationally Ready National Service
ORNSmen	Operationally Ready National Servicemen
OV	Organizational Value
PATUN	Perwira Penuntun (Facilitating Officers)
PCAP	Position Competencies Assessment Programs
PDF	Professional Development Framework
PETA	Pembela Tanah Air (Defenders of the Motherland)
PKB Kejuangan	Program Kegiatan Bersama Kejuangan (Joint Staff and Command Activities to instill Kejuangan values)

PMD	Professional Military Development
PME	Professional Military Education
PV	Personal Values
RNZAF	The Royal New Zealand Air Force
RNZN	The Royal New Zealand Navy
SAF	Singapore Armed Forces
SAF CLD	SAF Centre of Leadership Development
SAFTI	Singapore Armed Forces Training Institute
SAFTI-MI	SAFTI-Military Institute
SARS	Severe Acute Respiratory Syndrome
SESKOAD	Sekolah Staff dan Komando Angkatan Darat (Staff and Command School of the Army)
SESKOAL	Sekolah Staff dan Komando Angkatan Laut (Staff and Command School of the Navy)
SESKOAU	Sekolah Staff dan Komando Angkatan Udara (Staff and Command School of the Air Force)
SPERSAD	Staf Personel Angkatan Darat (Personnel Staff of the Army)
TNI	Tentara Nasional Indonesia (Indonesian National Armed Force)
TNI AD	Tentara Nasional Indonesia Angkatan Darat (Indonesian National Army)
UN	United Nations
US	United States
WWI	World War One
WWII	World War Two

INDEX

Index

Index

Index

Index

Index